ALT.FRACTALS

ALT.FRACTALS

A visual guide to fractal geometry and design

Eric Baird

Chocolate Tree Books

Published by Chocolate Tree Books, UK   www.chocolatetreebooks.com

Printed by Lightning Source

British Library Cataloguing in Publication Data.

A catalogue record for this book is available from the British Library.

Cataloguing data

Baird, Eric

    Alt.Fractals: A visual guide to fractal geometry and design

    First edition.

    232 pp. (i-vi, 1-225) size: 234 mm × 156 mm

    250+ numbered figures and illustrations (b&w). Includes index of figures.

    ISBN13:   978–0–9557068–3–7 (paperback)

    ISBN10:   0955706831

    1. Fractals, Geometry, Design

0-1021-122

Index

Graphics Tools:	
Main graphics:	CorelDraw®
IFS scripted graphics:	CorelDraw® + VBScript
Formula-based pixel/voxel graphics:	GFA BASIC, VB6.0
3D raytracing	Blender 3D

All trademarks are the property of their respective owners

1. Fractals

The word "fractal" was originally coined by Benoît Mandelbrot, to refer to what he described as "fractured" or "broken" geometry, partly because of what Mandelbrot referred to as "roughness" in the resulting shapes, and also perhaps partly because the new subject of fractals had gathered together examples that had previously been considered "pathological" or "broken" mathematics. Fractals broke some of the basic rules of classical geometry: Shapes could have infinite perimeter lengths and zero surface area or volume, or infinite detail, and the length of a line often depended on how closely you looked at it. Fractals seemed to represent a hypnotic combination of simplicity and intricate complexity, and often made you feel that if you just stared at them long enough that you'd grasp something deep about how the universe worked.

Like music, fractals contain repeating themes and details, sometimes with variations, and sometimes without, and fractal shapes typically contain an apparently infinite number of versions of themselves. Fractals come in different types and styles: They can branch, divide, cluster, speckle, scribble, clump, twist, spawn, flip, rotate, distort and replicate. They can have parts with any dimensionality embedded in or overlapping into any number of dimensions. They have recognisable features that repeat and recur across space and across scale.

So what *is* a fractal? It's a difficult question to answer. Some would say that the defining property of these shapes is **scale-dependent length** (the length of a side can depend on the size of ruler being used).

Another attempted definition could involve **fractional dimensionality**. If we scribble on a piece of paper for an infinite amount to time, so that the scribble ends up including every possible line that could be drawn on the paper, then the mess of one-dimensional line-segments would start to take on some of the properties of a two-dimensional plane. Some fractals can be assigned a **Hausdorff Number** to describe how strongly the shape intrudes into other dimensions, but Hausdorff numbers aren't the complete story, and with some shapes it's difficult to know how to compute them.

A third definition might involve **self-similarity** or **recursion** – we might say that a shape is fractal if shows similarity at all scales. Another involves **reducibility**: Fractal methods allow simple rules, applied repeatedly, to produce shapes and structures of arbitrary complexity.

Most people are probably familiar with fractals through the **Sierpinski Triangle** and **Pyramid**, the **Sierpinski Carpet** and **Menger Sponge**, the **Koch Snowflake**, and from the iconic **Mandelbrot Set** and **Julia Set** images, so we'll start with those.

1

2.   A few standards

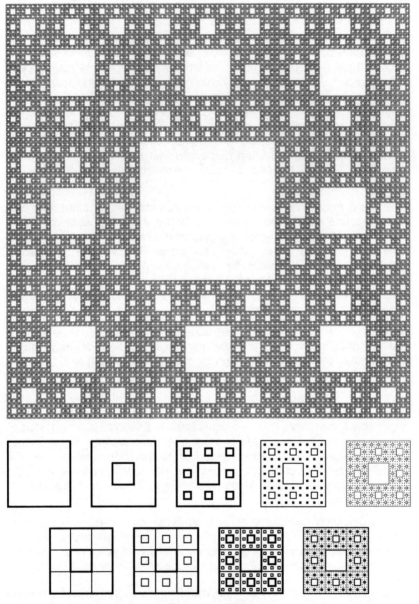

Figure 2-1: Standard Sierpinski Carpet

2. A few standards

The **Sierpinski Carpet** is one of the simplest fractals. We take a square, divide it into a 3×3 grid of smaller squares, and delete the grid's centre. We then take the eight remaining squares, divide *those* into a 3×3 grid, delete the centre, and repeat. The result is Figure 2-1.

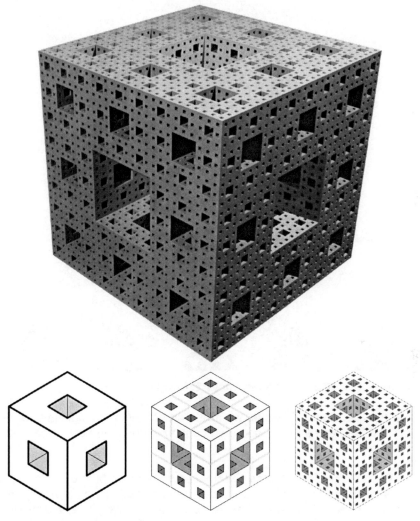

Figure 2-2: Menger Sponge

The Sierpinski Carpet extends neatly into three dimensions to give the **Menger Sponge**. We can divide each face of a cube into a 3×3 array, tunnel through the central block, and repeat. This gives a cubelike solid whose "shadow" is the Sierpinski Carpet for all six faces.

Figure 2-3: Standard Sierpinski Triangle

For the **Sierpinski Triangle**, we divide an equilateral triangle into four, discard the central piece, and then continue subdividing the remaining three pieces.

The Sierpinski Triangle and Pyramid (over) can be built by "additive" or "subtractive" methods: the shapes can be made by subdividing, or by clustering to produce larger identical copies. The Sierpinski Triangle can be made by grouping three triangles, then grouping three groups, and then grouping three *super*groups, and so on. The Sierpinski Pyramid uses clusters of four, the "carpet" groups of eight, and the Menger Sponge, clumps of twenty.

These shapes all look precisely the same at any scale.

2. A few standards

Each of these shapes can be used as a building block to construct larger copies.

Figure 2-4: Sierpinski Pyramid

The **Sierpinski Pyramid** is the extension of the Sierpinski Triangle into three dimensions. It's based on the **tetrahedron**, is a three-sided pyramid with a triangular base, giving a solid with four identical triangular faces.

We can make the outline of a larger tetrahedron by assembling four of these, corner-to corner, or we can divide a single tetrahedron into four corner pieces and discard the centre. As the number of stages (or *iterations*) increases, a builds a three-dimensional fractal shape develops.

Figure 2-5: Shadow of the Sierpinski Pyramid

The Sierpinski Pyramid isn't quite as simple as the Menger Sponge, in that although each of its four faces is identical to Figure 2-3, its "holes" don't extend through the shape in the same way.

Each *face* of the Sierpinski Pyramid is a Sierpinski Triangle, but **the solid's *shadow*** gives a different sort of repeating fractal shape that's already taking us into non-standard territory. It clearly repeats at different scales, and it's recognisably fractal, but the rules that generate it aren't clear from the final flat image, unless you already know how it was generated.

When we take shadows or cross-sections of standard fractals, other fractals are often only one step away.

2. A few standards

Figure 2-6: Koch Snowflake and Curve

The **Koch Snowflake** is another "standard" fractal. Starting with an equilateral triangle, we divide each face into three segments, and replace the centre segment with a smaller equilateral triangle. We then divide all the new line segments into three and replace the centre segments of those. And then we repeat, indefinitely.

The resulting Koch Snowflake is infinitely crinkly.

Alt.Fractals

2. A few standards

Figure 2-7: The Mandelbrot Set (two-page)

Alt.Fractals

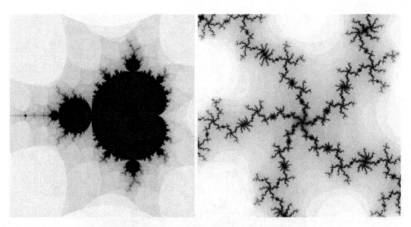

Figure 2-8: Mandelbrot Set, and detail

Benoît Mandelbrot discovered the **Mandelbrot Set** in the 1970s while studying the Julia Set. Unlike the previously-mentioned patterns, The Mandelbrot Set isn't generated by repeatedly adding or subtracting *shapes*, but by repeatedly applying a formula to test whether or not a point belongs within a particular category. We repeatedly apply a formula that keeps jumping our point to new locations, and check whether the point eventually escapes from the region altogether. If it stays trapped, it's counted as part of a "set" of trapped points.

It's traditional to colour the "trapped" points black, and to assign different colours to the "external" points depending on how many times the formula has to be applied before the equation successfully "escapes". This can produce psychedelic-looking effects. The boundary of the Mandelbrot has infinite detail, but its shapes are more organic-looking than with the constructional fractals like the Sierpinski Triangle, and show infinite subtle variation.

The swirling patterns that we see when we zoom in on the Mandelbrot set evolve and change character the deeper we get, but one of the strongest themes is the recurring appearance of tiny copies of the larger shape, as "islands" with dense surrounding structure.

Mandelbrot came up with the general name "fractals" and is credited with essentially inventing the subject as a legitimate field of study. Before Mandelbrot, some of these shapes were known and studied as individual mathematical curiosities, but there wasn't a general term to refer to all of them, and they were sometimes treated as examples of "bad" or "pathological" mathematics.

2. A few standards

Figure 2-9: Mandelbrot Set, more details

Julia Sets

The standard Mandelbrot equation only has two adjustable parameters, which are plotted on the horizontal and vertical axes of the page to draw the shape. **Julia Sets** (named after **Gaston Julia**) have four. Normally we use two of these to draw an image, and the other two as a sort of index to decide which version or variation we'll be plotting. Since we can specify the two "index" values with arbitrarily precision, there are technically an infinite number of Julia set images.

Like the Mandelbrot set, Julia sets have organic shapes and recurring themes at different scales.

The Mandelbrot and Julia sets both use the same basic repeated equation,

$$z \rightarrow z^2 + c$$

Each time we reapply the equation, we take the previous value for **z**, square it, and add a fixed constant. To make the resulting shapes more interesting, we use "**complex numbers**" for **z** and **c**, that each have two independent components that can multiply together in odd ways. It can be handy to label the two components of the start value of **z** as "**A**" and "**B**", and the two components of the thing we add each time as "**C**" and "**D**". The Mandelbrot Set starts the process off with **z=0** (**A=0**, **B=0**), leaving us with just **C** and **D** as variables, and plotting **(0, 0, C, D)** gives us the familiar Mandelbrot shape.

On the other hand, Julia Set images are traditionally generated by keeping the constant's values (**C&D**) fixed, and graphing how the start value of **z** affects the outcome, instead. The full Julia Set is four-dimensional (**A B C D**), and includes the Mandelbrot Set as a special case: the usual Julia Set images are cross-sections through this larger 4D shape parallel to the plane **AB**, and the Mandelbrot Set is a special slice that runs through the exact centre of the same 4D shape, on the **CD** plane (see section 13). The standard Julia images are tabled in Figure 18-1.

We can also try generating Julia Sets for other functions, such as

$$z \rightarrow z^3 + c$$

, or

$$z \rightarrow SIN(z) + c$$

, and these will all have their own Mandelbrot Set counterparts.

2. A few standards

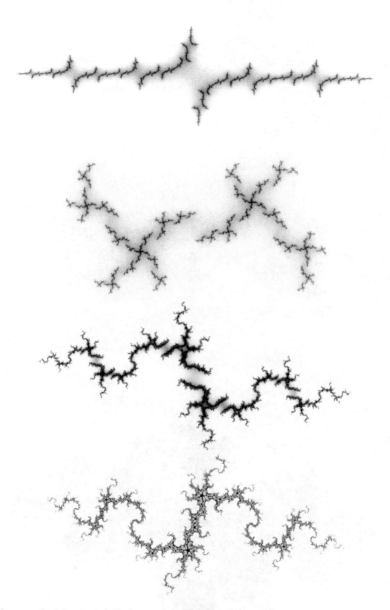

Figure 2-10: A selection of Julia Set images

Alt.Fractals

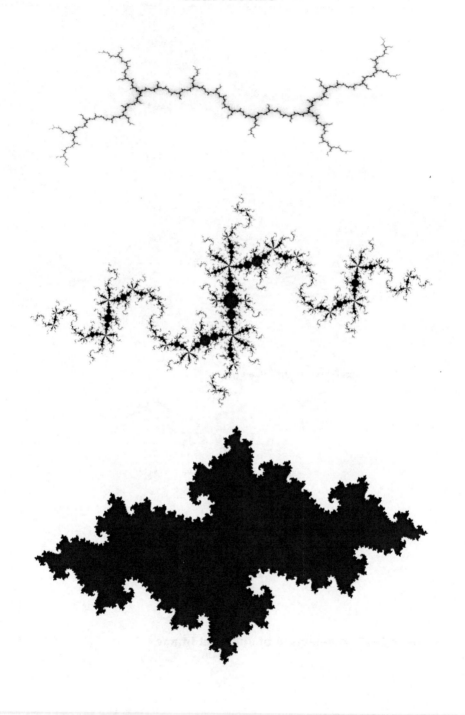

2. A few standards

Figure 2-11: More Julia Set images and "zoomed" details

Figure 2-12: More Julia Set details

3. Not the Koch snowflake

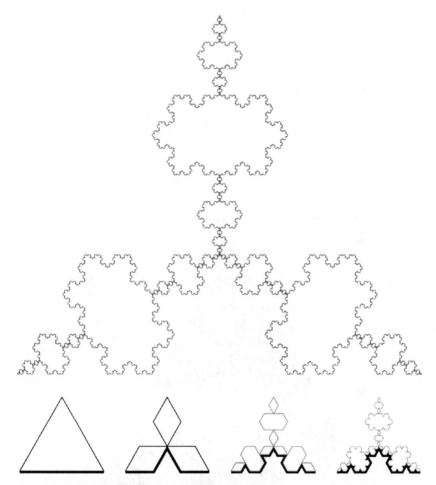

Figure 3-1: Negative Koch Snowflake

The standard Koch Snowflake (Figure 2-6) repeatedly adds third-scale triangles to every free piece of perimeter line. If we subtract the triangles instead of adding them, we get the **Negative Koch Snowflake** in Figure 3-1, whose outline consists of three inward-pointing **Koch curves**.

The remaining solid parts of the shape can be broken into three identical shapes each outlined by a mirrored pair of Koch curves (Figure 3-2).

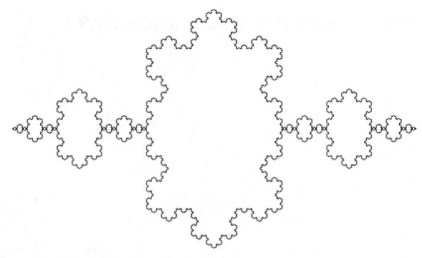

Figure 3-2: Koch Curve, mirrored

Figure 3-3: "Edge-Packed" Negative Koch

Since the negative Koch has an overall triangular outline, we can fit third-sized copies into its edge-gaps to produce Figure 3-3. All these shapes have a recurring "double-snowflake" theme.

3. Not the Koch snowflake

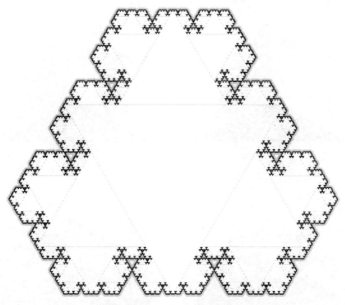

Figure 3-4: Koch with daughter-scaling increased to 50%

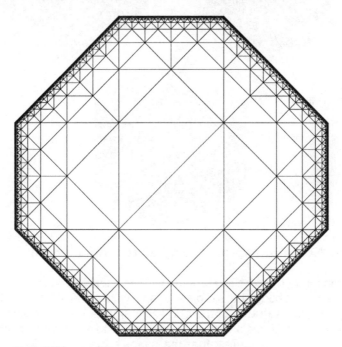

Figure 3-5: Tiling with right-angled triangles

Figure 3-6: Koch Approach, applied to a tetrahedron

It's tempting to try to generate a three-dimensional version of the Koch Snowflake by taking the simplest shape with a triangular profile – a triangular-faced **tetrahedron** – and fixing smaller copies to each of its four faces. The shape's profile initially turns into a six-pointed star, but additional iterations merely turn the shape's external surface into an ever-closer approximation of a cube.

3. Not the Koch snowflake

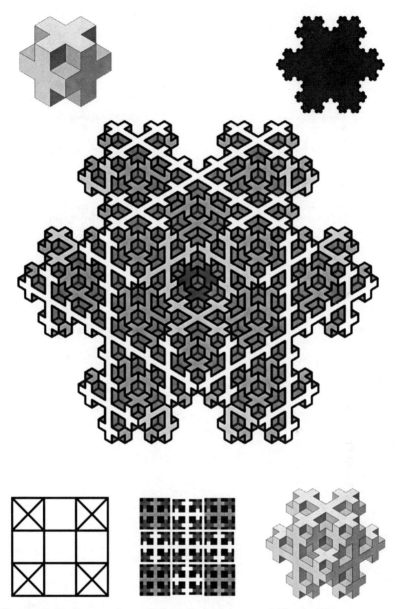

Figure 3-7: Cornerless Cube-Stacking, and Koch-like silhouette

The solid shape that *does* generate a Koch-like profile turns out to be the cube – repeatedly subdividing a cube into 3×3×3 blocks and removing the eight corner cubes generates Figure 3-7.

Figure 3-8: "Squares" counterpart to the Koch Snowflake

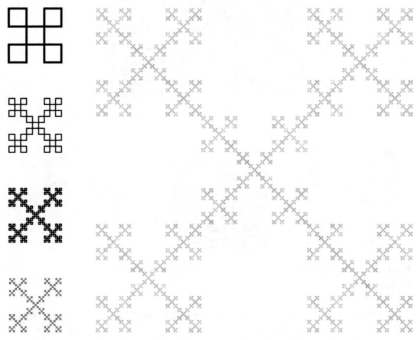

Figure 3-9: "Negative" version of the "Squares" Koch

3. Not the Koch snowflake

Figure 3-8 uses third-sized squares instead of the Koch Snowflake's triangles, giving a square outline rotated through 45 degrees. The "inverse" version (Figure 3-9) has an area that shrinks towards zero after an arbitrarily-high number of operations.

Although this version is built from squares, after a sufficient number of iterations it starts to look like a branching linear fractal

Figure 3-10: Everted Menger Sponge, and profile

The "**Everted Menger sponge**" incorporates both fractals.

As with the "proper" Menger Sponge, we start with a cube and repeatedly divide each face into nine squares – but instead of punching a hole through the solid for each centre square, we add a cube to the outside. This produces a loose "stepped pyramid"-type structure on each face, with an arbitrarily thin stream of cube-steps forming a line linking each face-centred peak.

Viewed face-on, the shape's shadow corresponds to Figure 3-8's "Square Koch".

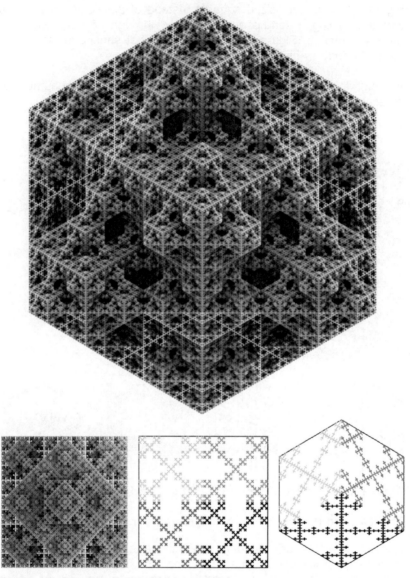

Figure 3-11: Everted Menger Sponge – crosswork

As well as generating the "square Koch", this solid also produces the "anti"-version. It's ridges produce a set of pseudosurfaces, and the "stepped pyramid" structures stop it from resting on its original faces – if dipped in ink and placed on a sheet of paper, the resulting inky "footprint" will be the fractal series of crosses from Figure 3-9.

3. Not the Koch snowflake

Figure 3-12: Thirds-Square sequence

For a more crinkly outline, we can add two squares per face, and align them with the corners. Extended into three dimensions, this gives a more "blobby" surface.

Figure 3-13: Koch Snowflake profile, using hexagons

Swapping the Koch's third-scale triangles with hexagons doesn't obviously change the resulting "Koch-y" profile.

Both methods of making the "snowflake" outline (triangle-based and hexagon-based) have their own advantages: the standard approach can claim superiority for using a simpler three-sided building block, while the "hexagon" approach has the advantage that its building-blocks already have the same sixfold symmetry as the final shape.

3. Not the Koch snowflake

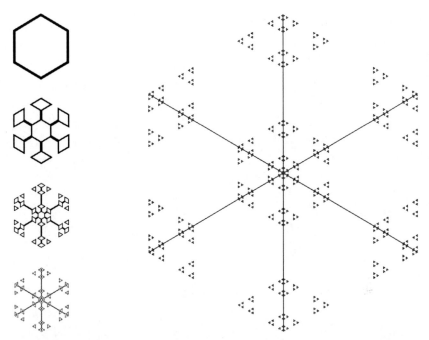

Figure 3-14: Removing third-scale face-centred hexagons

On the other hand, if we repeatedly *remove* hexagons (ignoring overlaps), the shape ends up with zero area. The shape shows three fractal styles: the central hexagon always reduces to a smaller copy of the global shape, of a hexagon linked to six diamonds, the diamonds reduce to a pair of linked diamonds and two floating triangles, and the triangles reduce to triangle-triplets, and, ultimately, to triangular "dust".

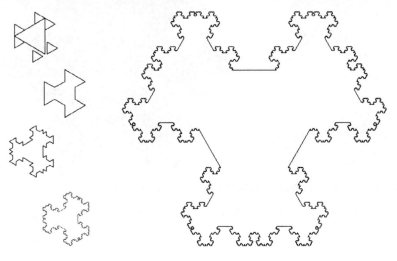

Figure 3-15: Spikeflake

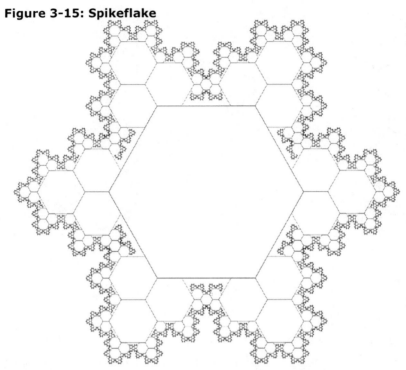

Figure 3-16: Hexagonal Corner-Cluster Snowflake

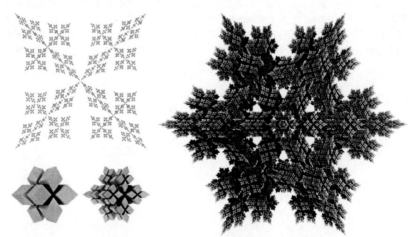

Figure 3-17: Teased Square and Teased Cube

The Koch Curve and other similar shapes can also be created using a triangle-subdivision method, where solid triangles are repeatedly split into two smaller copies and a wedge-shaped gap (Figure 3-18).

3. Not the Koch snowflake

Figure 3-18: Subdividing triangles: 30-30-120 degrees

Figure 3-19: Subdividing triangles: 42-42-96 degrees

Starting with a symmetrical 30-30-120-degree triangle, the resulting "gap" has an angle of sixty degrees, and the process ends up creating three interlocking Koch curves, one for each side of the original triangle. Figure 3-19 shows the result of using an initial triangle with 42-degree corners, giving a more spiky "skinny Koch" curve. This general pattern also appears in the Pentangle Briar (Figure 30-3), and on each of the six faces of the fractal Delta solid in Figure 33-5.

4.   Not the Sierpinski Triangle

Figure 4-1: Corner-Subtracted Cube, and imprint map

The Sierpinski Triangle is a recurring theme in many "geometrical" fractals, and variations sometimes appear in unexpected places.

Figure 4-1 shows the result of repeatedly removing all the protruding corners of a solid that starts out as a cube. Every time a corner is removed, three more appear. The result is similar to the Sierpinski Triangle, but with different spacings (see also Figure 10-4)

4. Not the Sierpinski Triangle

Figure 4-2: Simple Triangular Carpet

The Sierpinski Triangle is probably the simplest and most elegant triangular fractal, but there are others. Comparing the Sierpinski Triangle and Carpet, we find that the Triangle uses corner-connected shapes, while the Carpet uses more solid edge-to-edge tiling.

Using the chunky "carpet" approach with equilateral triangles gives the more blocky shape in Figure 4-2.

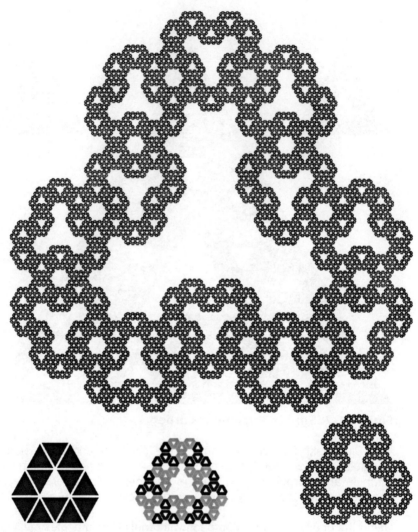

Figure 4-3: Decornered Triangular Ring

Although Figure 4-2 isn't especially interesting, it can be used as a jumping-off point for a number of more elegant fractals. We don't need the three outmost corner pieces to build a solid ring around the central triangular space, and leaving these out gives the more funky shape in Figure 4-3.

4. Not the Sierpinski Triangle

Figure 4-4: "Starry" Triangular Tiling

Instead of dividing the major profile of Figure 4-2 into fifteen identically-sized triangular tiles, we can use just six (three large and three small).

This gives a more visually pleasing effect that looks less like a simple grid-array of holes, and feels more "random", although it's still very much an ordered pattern with a strict grid.

Using multiple scales within a template is often a good way of creating fractals that look less "regimented", but can makes the construction process more difficult.

Figure 4-5: Floating Islands

Figure 4-5 surrounds each triangular tile with three half-size triangular "islands" to give a larger overall triangular profile.

The end-result is similar to the "Dragon" fractals that we'll meet in section 35.

4. Not the Sierpinski Triangle

Figure 4-6: Fractal Labyrinth

Adding corner-triangles to Figure 4-5 gives an interlocking series of fractal islands, and a labyrinthlike path running between them.

Figure 4-7: Mitsubishi Tilings (see also Figure 39-4)

"**Mitsubishi**" is Japanese for "Three Diamonds" (mitsu, hishi), and the "three diamonds" symbol is used by the Mitsubishi Group as a company logo.

The "three-diamond" shape can be subdivided fractally, but can also be used to produce a number of interlocking tiling systems that can also be used as the basis of fractal patterns.

4. Not the Sierpinski Triangle

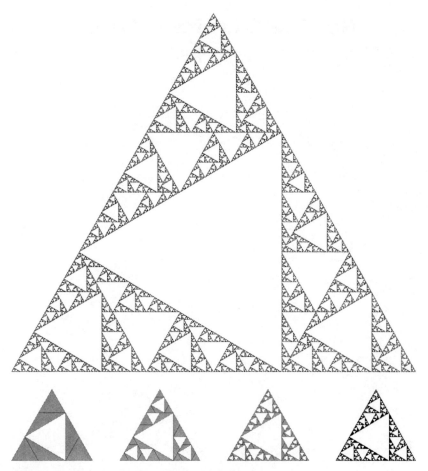

Figure 4-8: Triangular Crazy Tiling

Figure 4-8 might look quite chaotic at first glance, but it uses a tightly ordered tiling scheme – the mix of sixty- and ninety-degree angles confuses the human eye when it's searching for signs of reducible order.

The large triangle has three equilateral triangular tiles cut from its corners, leaving a hexagon. This then has three wedges removed, leaving a central triangular void that (somewhat incongruously) is now at ninety degrees to the original shape. Each regular triangle is then divided like its parent, and each wedge is cut into a pair of equilateral triangles and a pair of smaller wedges.

5.   Not the Sierpinski Carpet

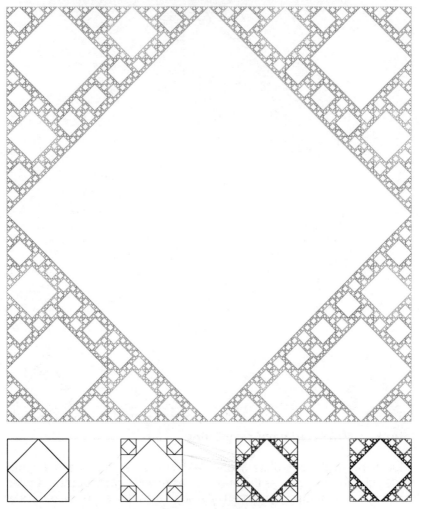

Figure 5-1: Not the Sierpinski Carpet (#1)

The Sierpinski Carpet isn't the only recursive way to puncture a square with square holes. Borrowing the "corner triangles surrounding a central void" theme from the Sierpinski Triangle, and applying it to a square, gives the shape in Figure 5-1.

Each triangular corner-piece breaks down into a square (a smaller clone of the entire figure) and two more corner-pieces. It takes an infinite number of squares to tile a corner, even before subdivision

5. Not the Sierpinski Carpet

Figure 5-2: Not the Sierpinski Carpet (#2)

The triangular corner-pieces in Figure 5-1 are each built from an infinite number of square tiles of cascading sizes. This is more explicit in the different proportions of Figure 5-2.

It's a little more sophisticated/complicated than the usual carpet, and it doesn't led itself so easily to 3D structures (although it could be used as a template for cube-based or octahedral foam). The result is less regimented and more satisfyingly "foamy" than the usual version.

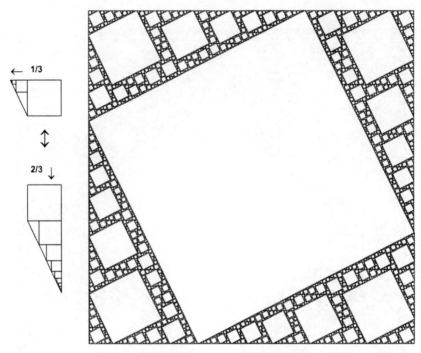

Figure 5-3: Tiling with 1/3 & 2/3 -ratio sequences of squares

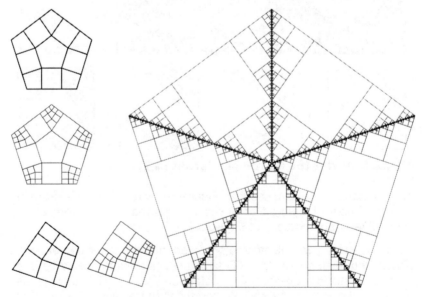

Figure 5-4: Square-tiling a non-rectangular shape

5. Not the Sierpinski Carpet

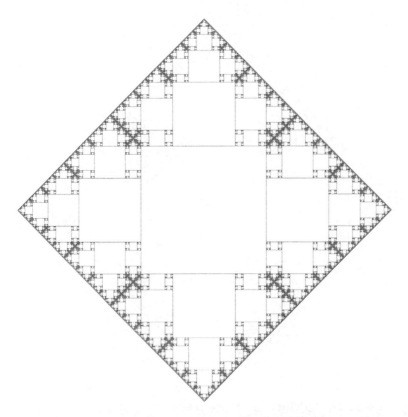

Figure 5-5: Square Face-Centred Tiling (50% scale ratio)

Figure 5-6: Square Corner-Connected Tiling (50% scale ratio)

Figure 5-7: Circle-Tiling at 90 degrees

Figure 5-7 is a redrawn version of Figure 5-5, with the original squares replaced by circles. The perimeter profile and network map of where each shape touches its neighbours are identical to the original.

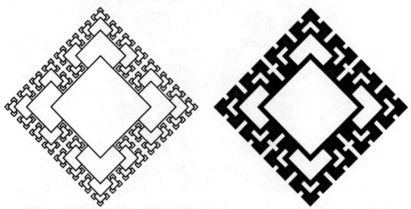

Figure 5-8: Overlapped half-size squares

5. Not the Sierpinski Carpet

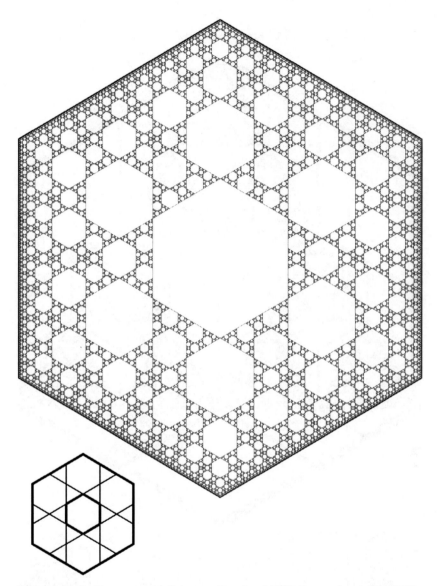

Figure 5-9: Hexagonal Corner-Contact Tiling, scale ratio 50%

In this hex carpet, we start with a single hexagon and repeatedly add half-size copies to every free corner in the figure. After an infinite number of iterations, the full carpet (ignoring overlaps) produces a solid hexagon with an area exactly nine times that of the original.

The shape also works as the two-dimensional shadow of a 3D array of half-size-sequence cubes, connected by their corners.

6.   More Hexagons

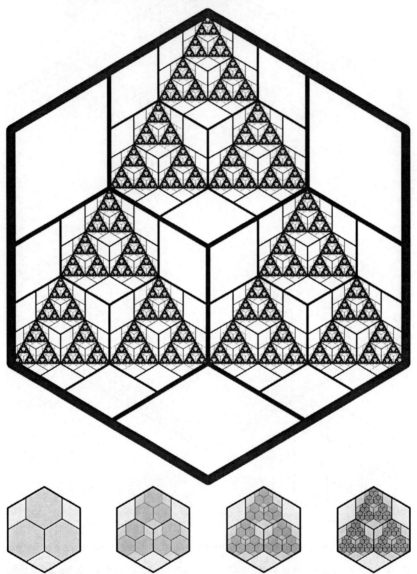

Figure 6-1: Hexagon Triplets – Sierpinski Triangle

6. More Hexagons

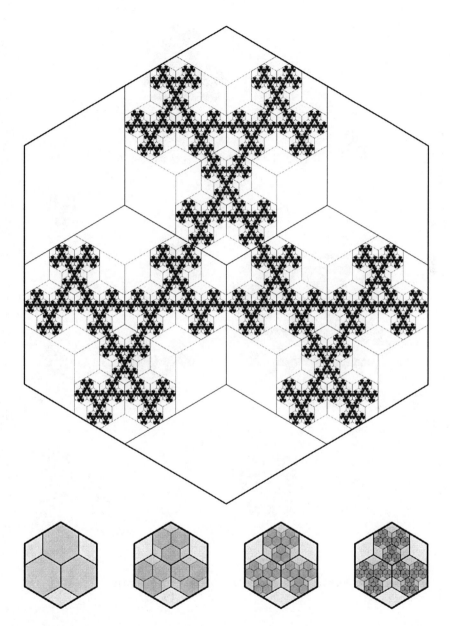

Figure 6-2: Hexagon Triplets (alignment #2)

Figure 6-3: Some Different Hex-Clustering Modes

6. More Hexagons

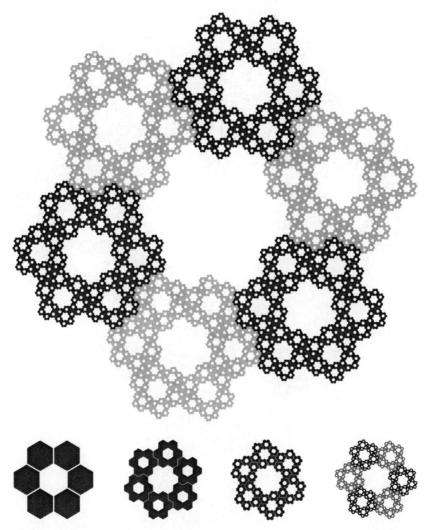

Figure 6-4: "Snug-fit" Hex Ring

Adding three triangles around a triangular hole, or eight squares around a square hole, gives a larger shape whose perimeter is a perfect copy of its smaller components.

This isn't true of the **Hex Ring** – its perimeter becomes progressively more complex and crinkly at each iteration (like the Koch Snowflake).

Because the Hex Ring has more than one face per "side", there are different options for how adjacent shapes can touch and/or interlock.

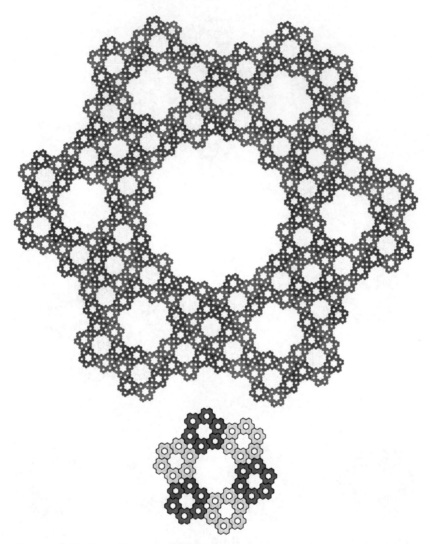

Figure 6-5: Hex Ring variations

With the "snug" scheme, the perimeters of the smaller shapes tile exactly, with no gaps, and it'd seem that we have no free choices as to how to construct it ... but the tiling scheme is "chiral", and comes in left-handed and right-handed versions.

There are also more "gappy" tiling laws that can snap the shapes together in different ways at different iterations, as "best fit" quantised approximations of a deeper underlying rule.

7. Booleans and Transparencies

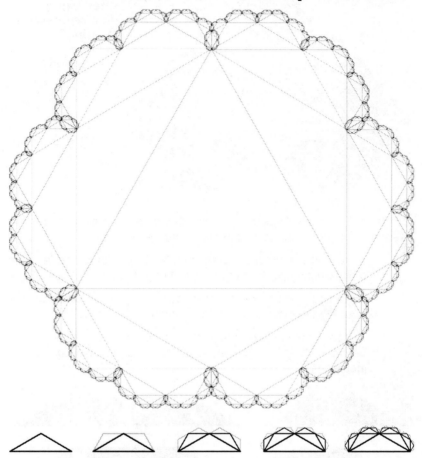

Figure 7-1: Wedge Fractal

The wedge fractal in Figure 7-1 introduces the problem of how to deal with overlaps – the "islands" inside the figure's edge represent overlapping regions, and there's no obvious single answer for how these should be drawn.

Designers and mathematicians with a background in tiling problems will tend to leave out any overlapping parts, and consider that when a new tile's edge butts cleanly up against another, that the new edge is a "dead end" and plays no further part in the process. Programmers with a background in IFS (section 34) will tend to include overlapping features, since overlap-testing isn't part of their approach.

Alt.Fractals

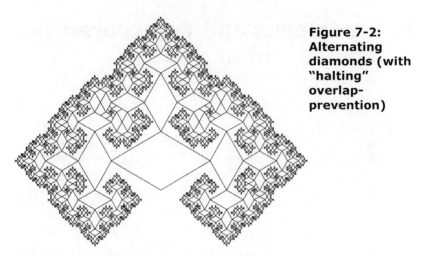

Figure 7-2: Alternating diamonds (with "halting" overlap-prevention)

Overlapped line-diagrams can quickly become confusing, and there seem to be three main ways of dealing with them:

1. … to halt the process when further detail would overlap with other existing parts of the diagram (e.g. Figure 7-2),

2. … to use transparencies or colour-coding to make a region's intensity or colour vary with the number of overlaps, or

3. … the "Boolean" approach, for the colour of a region to "flip" between black and white each time its overlap count increases.

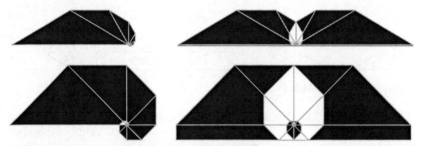

Figure 7-3: Two Common Overlapping Spiral-pairs

Finally, if we *are* to be allowing overlaps, we have to consider whether a newly-drawn component should be allowed to use *all* of its corners and faces for creating more offspring, or just the ones that aren't already connected to its parent. The Koch Snowflake's construction implicitly assumes that when we add triangles to a parent object, only the "free" edges are available for further construction. If we ignore this convention, we can get a further set of fractal variations.

7. Booleans and Transparencies

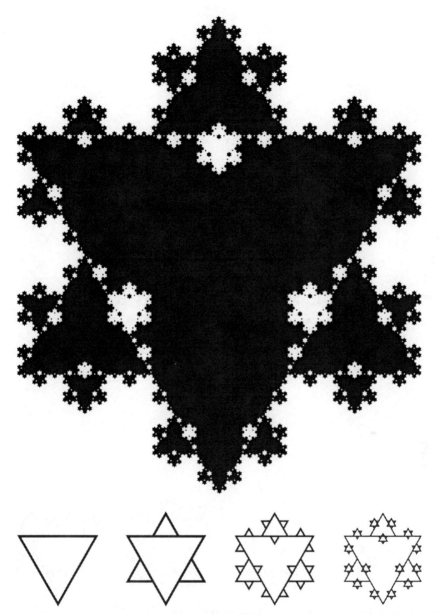

Figure 7-4: "Boolean" Koch Snowflake

Allowing each component triangle in the Koch snowflake to spawn offspring on *all three* of its faces gives Figure 7-4.

Figure 7-5 is the "Boolean" version of the hexagon-based snowflake variant from Figure 3-13.

Figure 7-5: Boolean "Hexagonal Koch"

These alternative versions of the Koch have the same outline, but contain snowflake archipelagos: cascading series of identical copies of the full shape. While the "normal" Koch is a self-contained island that only reveals more detail when we zoom *in*, the "Boolean" versions

7. Booleans and Transparencies

also let us zoom *out* to show optional external copies alongside and enclosing the main figure, while still obeying the same template. Allowing *internal* structure allows corresponding *external* structure that lets the shape extend and scale outwards in a similar way to the Sierpinski Triangle and Carpet – the "modified" snowflakes no longer have to represent fully-closed systems.

Figure 7-6: "Boolean" Square-tiling

Figure 7-7: A "Boolean" version of the Figure 2-5 Sierpinski Pyramid shadow

Figure 7-8: "Boolean" vs. "transparency" Renderings

8.   Not the Sierpinski Pyramid

Figure 8-1: "Sierpinski Octahedron"

The Sierpinski Pyramid has four tetrahedra tucked into the corners of a larger tetrahedron. In the octahedral version, the smaller shapes share some of their edges as well as their corners. All six smaller copies now touch at the centre of their parent shape, at a single point, and there's no central void.

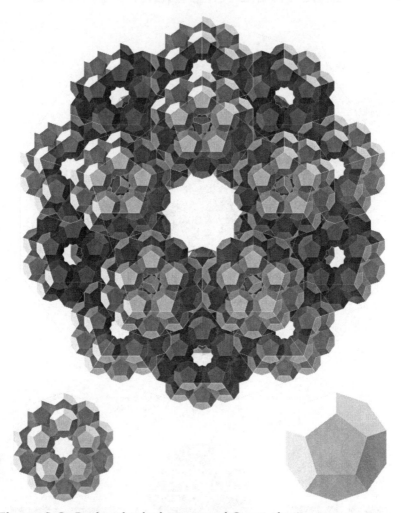

Figure 8-2: Dodecahedral-cornered Supercluster

Other regular and "semiregular" solids can be edge-connected in a similar way to produce self-similar fractal clusters. The **dodecahedron** has twelve faces and twenty corners – we can fit a smaller dodecahedron into each corner of the original solid so that the smaller shapes just touch, edge-to-edge, to get a dodecahedral cluster.

Each smaller building block can then be replaced by another smaller set of twenty. This process can be repeated indefinitely.

8. Not the Sierpinski Pyramid

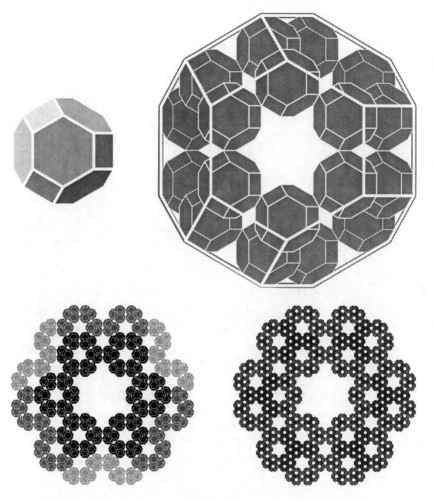

Figure 8-3: 14-sided Polygonal Clusters, and Shadow

Although there's not a hexagonal regular solid that's equivalent to Figure 8-2, the 14-sided shape used in Figure 8-3 comes close.

This is a **truncated octahedron**, with eight hexagonal faces and six square faces. It can be thought of the result of lopping the six corners off an octahedron so that the shape's eight original triangular faces become hexagonal, or as the result of aggressively cutting the eight corners off a cube, and cutting so deeply that the eight new faces, which would otherwise have been triangular, cut off each others' corners and turn into hexagons.

Figure 8-4: Corner-Connected Octahedra

Simple "corner-contact" fractals are comparatively easy to model in computer software, because the points of the parent shape can be simply extrapolated or projected through a corner shared with its daughter shape (with a scaling factor), to give all the coordinates for the next shape in the sequence.

This makes it easier to extrapolate a parent shape to its offspring without resorting to nasty trigonometry, even when the shapes are irregular. The method also works easily in any number of dimensions.

8. Not the Sierpinski Pyramid

**Figure 8-5: Corner-Connected Tetrahedra
(wireframe, raytraced, and shadow)**

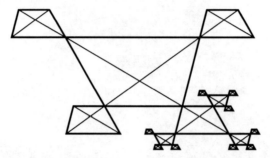

Figure 8-6: Corner-extrapolating an irregular shape

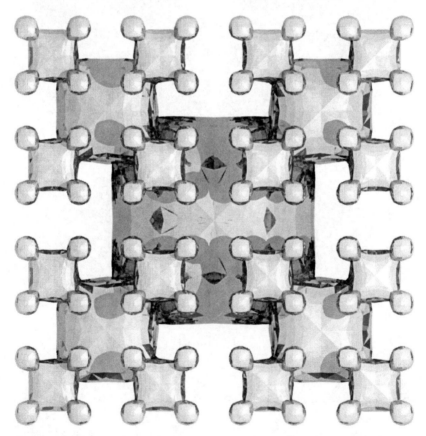

Figure 8-7: Corner-connected Cubes (overlapped and smoothed)

9.   Not the Menger Sponge

Figure 9-1: " 1,0,0,1" Counterpart to the Menger Sponge

This variation on the standard Menger Sponge might look like a trivial variation, but it has some properties that the normal sponge lacks – it works better for binary division, it's a space-filling solid, and it's related to some other space-filling fractal shapes (section 40).

The solid's volume halves at each iteration, so that it ultimately tends to zero.

Figure 9-2: The "Snowflake" counterpart to the Menger Sponge

Figure 3-7 showed that we could repeatedly subdivide a cube using a 3×3×3 grid and discard the corner cubes to get a snowflake silhouette.

If we also remove the "centre" and "centre-face" cubes, the corner view gives the same perimeter silhouette as before, but punctured by an infinite series of snowflake-shaped holes.

9. Not the Menger Sponge

Figure 9-3: Snowflake Solid "shadow"

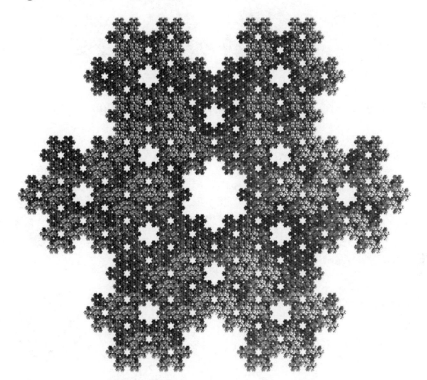

Figure 9-4: A 3D Model of the "Snowflake" Solid

Each iteration punches a fresh set of hexagonal holes into the shadow, and further iterations turn each hexagonal window into an increasingly crinkly star whose profile becomes more Koch-like as the number of iterations progresses. Figure 9-4 and Figure 9-5 show views of a 3D model of the solid built from more than a million facets, including the "Snowflake" and "Menger" profiles.

63

Alt.Fractals

The "snowflake-within-snowflake" view is visible from any of the eight corners, but the "shadow view" as seen from any of the six *faces* is a standard Sierpinski carpet, just as if we were looking at a Menger Sponge. The solid also has some interesting profiles at other angles:

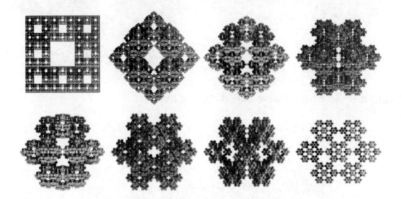

Figure 9-5: Alternate tilted views of the snowflake solid

So far, we haven't shown the **octagonal ring fractal**.

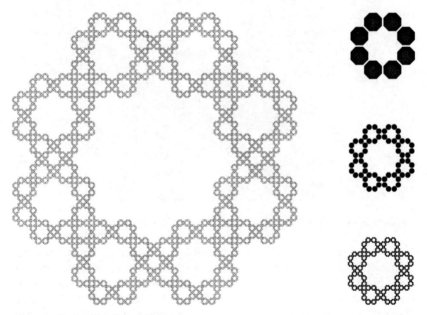

Figure 9-6: Octagonal ring

9. Not the Menger Sponge

Figure 9-7: Rhombicuboctahedron Clustering, and shadow

The octagonal rings may not be especially interesting in themselves, but they also turn out to represent a cross-section through a hollow **rhombicuboctahedron** cluster, which is a more interesting creature.

Alt.Fractals

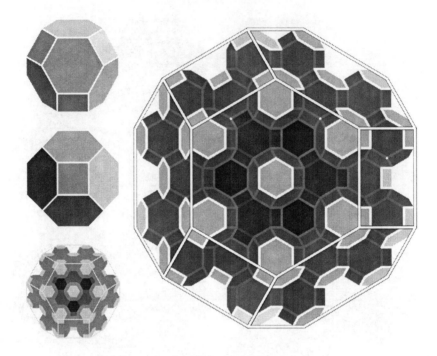

Figure 9-8: Solid 14-sided polyhedral clusters

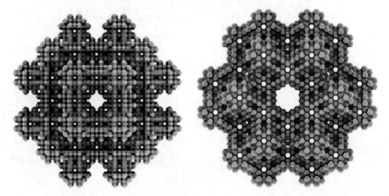

Figure 9-9: Hollow 14-sided polyhedral clusters (two views)

Because the 14-sided solid used in Figure 8-3 is a space-filling solid, it can also be "stacked" to build fractal clusters that are face-aligned.

The solid self-similar cluster in Figure 9-8 has the odd property that the solid effectively returns to its original shape after an infinite number of iterations. The "pierced" version in Figure 9-9 doesn't include centre blocks.

10. Negative Solids

Figure 10-1: Negative Menger Sponge

When dealing with fractal solids that have a lot of internal structure it can sometimes be useful to map out the voids within the structure rather than the solid parts. This "plaster cast" approach lets us see surfaces (and sometimes relationships) that would otherwise be hidden or difficult to see. And, of course, the "imprint" of one fractal tends to be another fractal.

Figure 10-2: Negative Sierpinski Pyramid

The standard Sierpinski Pyramid consists of a tetrahedral shape divided into four smaller tetrahedra and a central void. The central space has eight identical faces – four of these faces are shared with the corner pyramids, the other four are unused and normally act as "windows" that let us see into the Sierpinski Pyramid interior.

But we can also consider the Pyramid as a fractal network of octahedral spaces that completely fill the tetrahedral volume. Since many of the tetrahedra in the pyramid are adjacent to an external edge, the "negative" of the pyramid, which shows the octahedra as solid and the tetrahedra as space, gives a useful alternative view.

10. Negative Solids

Figure 10-3: Negative "Sierpinski Octahedron"

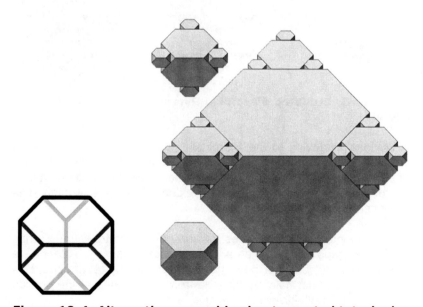

Figure 10-4: Alternative pyramid using truncated tetrahedra

69

Figure 10-5: Stellated Sierpinski Pyramid: Exterior and Interior

An obvious extension to the Sierpinski Pyramid is to add additional pyramids to the remaining sides of the central void to give a "star" shape. This variation isn't common because the exterior tends towards a simple cube (Figure 3-6), and because the interesting internal structure is hidden.

Plotting the internal spaces gives an array of octahedra with recurring "gaps". The gaps turn out to be additional octahedral spaces that were originally outside, but that have been "boxed in" by tetrahedra. This subtlety wouldn't normally be obvious from just looking at the outside of the conventionally-drawn shape.

11. The Fractal Library

Figure 11-1: "Fractal Library" Pseudocarpet

The floorplan of a library has to meet some special requirements. There's usually a broad entrance leading to a large atrium with the information desks. The large central space is well lit and provides light for the rest of the building. Branching away from the main space are smaller spaces for study areas, and then finally there are the shelves. The layout requires the maximum amount of shelving to be accessible along paths that are as short and wide as possible.

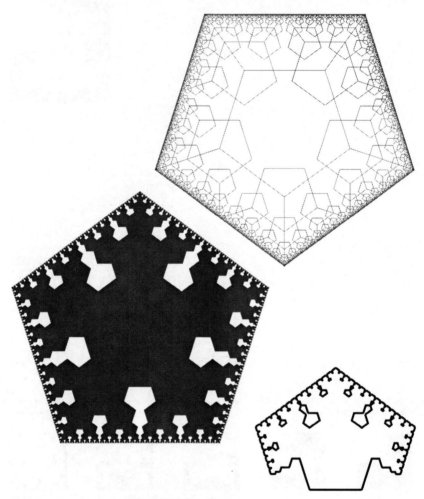

Figure 11-2: Overlapping Pentagons

Library book-classification systems like "Dewey Decimal" also tend to organise books into a single sequence, so and ideally the shelf plan would give a single uninterrupted line of shelving.

It's a fractal problem. Although modern libraries now tend to arrange their shelves in an ad-hoc grid of rectangular shelf "islands", some older libraries are designed around more sophisticated fractal floorplans.

11. The Fractal Library

Figure 11-3: Alternative renderings of a "Hex" floorplan

Figure 11-4: ... using triangles

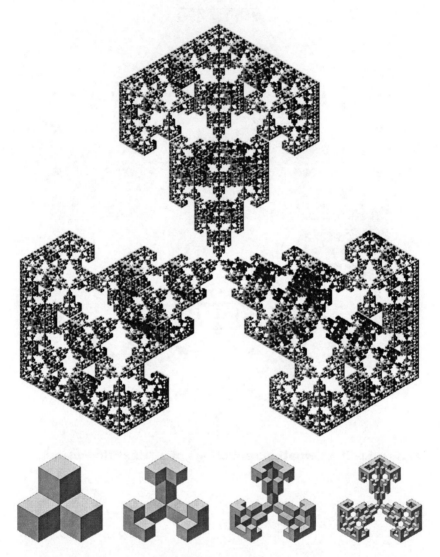

Figure 11-5: "Library Fractal" Pseudosponge (Fig 11.1 extended into three dimensions)

12.   Inside the Mandelbrot Set

The Mandelbrot and Julia shapes are essentially escape limits.

Suppose that we drop a pea somewhere above a drinking-glass. If the pea falls inside the glass rim, it's trapped. If it falls outside, it escapes. Plotting a graph of the positions that the pea was dropped from, and colouring each point of the graph according to whether the pea was captured or lost gives us a simple filled circle representing the set of trapped points. This is similar to the circular plot that we get when we plot the Julia set for **C=0**, **D=0** (Figure 13-3).

What happens if the pea lands on the edge of the glass? If it hits fractionally inside the rim's peak it might bounce and land inside the glass, but it hits slightly further in, (and is dropped from a sufficient height), it might bounce clear over the glass and escape, giving a thin white internal ring of additional escape positions within our black disc-shaped graph of captured points.

By the edge of this ring there should be a second ring where the pea bounces twice before capture or escape, and perhaps an unlimited series of further rings representing locations where the pea (theoretically) bounces a hundred or a thousand or a million times. And the pea might not be spherical. If it's cube-shaped and dropped corner-first, then the graph won't be a perfect circle, it'll have threefold symmetry (fourfold symmetry if it's dropped "flat"). The way that the cube will spin between bounces will then give different orientations to successive landing-points, making parts of the graph extremely complex at smaller scales. The "interesting" part of our graph can be broadened by increasing the size of the dropped object with respect to the size of the glass, so that captures are more difficult and multiple bounces more common. Each additional independent property adds another dimension to the model, and we can easily arrive at an eight- or ten-dimensional description of different factors that might affect where the pea eventually ends up.

With the Mandelbrot and Julia sets, the equation that we apply is comparable to the hypothetical equation that would tell us where the pea lands, given a known starting position and associated parameters. Each time the pea bounces, the parameters change for the next bounce. With the Mandelbrot and Julia sets, each iteration of the equation creates a new start position until the point escapes or until we give up waiting for it to get free (which might be a hundred or a thousand or a billion iterations).

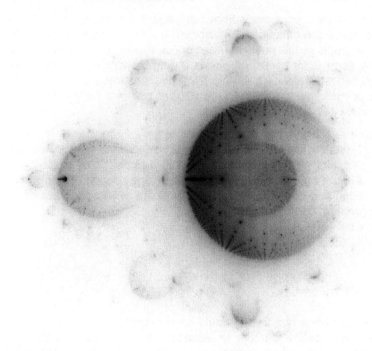

Figure 12-1: Mandelbrot attractors, for iterations 1-32

Tracking the "bounce points" for the first thirty-two iterations of the Mandelbrot set equation gives Figure 12-1, with Figure 12-2 showing the distribution for individual stages. Given an initially-even spread of points (Figure 12-2, top left), each iteration produces a progressively more complex set of focal points or attractors, with the odd- and even-numbered images alternating between shapes that have a leftmost lobe and shapes that don't. As the number of iterations increases, the growing "posthole" pattern of attractors makes paths and orbits through the set's region progressively more complex.

Although we aren't normally shown these paths, they can be used to plot further derivatives of the Mandelbrot Set (such as Melinda Green's "**Buddhabrot**"). Logically, each location that an escaping point lands on must also be an escaping point, and each location landed on by a point that counts as "trapped", must be another trapped point.

12. Inside the Mandelbrot Set

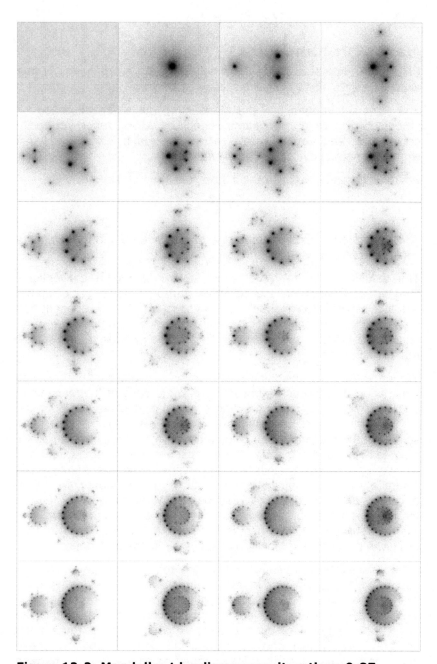

Figure 12-2: Mandelbrot landing zones, iterations 0-27

13.   The Julia Set in 4D

Mandelbrot and Julia

Before we attempt to think about how the Julia set behaves in four dimensions, it's worth revisiting some of what we've already learnt.

The Julia set images and the Mandelbrot image are all based on repeatedly applying the same equation to a range of point coordinates, and asking whether or not the result "escapes" after a certain number of cycles, or "iterations".

For the standard Mandelbrot and Julia sets, the equation used is

$$z \rightarrow z^2 + c$$

– we repeatedly take the current value of *z*, square it, and then add a constant (*c*) to the result to find its next value.

The results would be rather predictable if *z* was a conventional number, so instead we use **complex numbers**. A complex number has two separate components, a conventional "**real**" component that exists on the normal number scale, and an additional "**imaginary**" component that is a multiple of the impossible number "*i*", the square root of minus one.

We know the rules for doing multiplication and other operations with these special compound numbers, and in the case of the Mandelbrot and Julia sets, they produce chaotic and unstable results. Squaring the "real" part of a complex number always gives a positive result, and squaring the "imaginary" part gives a negative real outcome, so as we repeatedly apply the formula, the two parts of the numbers fight and wrestle against each other. Points originating in the Mandelbrot's central zone are obviously trapped, and any points further away than a critical distance obviously escape. Between these two regions there's a more ambiguous region, where the boundary twists and writhes and squirms more intensely as we look more closely … and this gives the Julia and Mandelbrot sets their fractal boundaries.

As we said in the previous section, the route that a point takes to escape isn't usually plotted in these images, only the "yes/no" result of whether the escape attempt eventually succeeds or not from the given position, and perhaps also some sort of colour-code or shading to suggest how many iterations the escaping points needed. Parts of the set are infinitely thin and don't show up well on plots, so these shadings give useful contour lines around detail that would otherwise be too small to see. These bands can be coloured in to give gradient or psychedelic effects.

13. The Julia Set in 4D

Four Parts

Since c is a complex number, and it's repeatedly being added to the z^2 term, z has to be as a complex number, as well. This gives us a total of four separate parts for the equation at any moment, two for z, and another two for c. If z starts with a value of zero, then the only adjustable parameters are the real and imaginary parts of c, but if we allow z to start anywhere, we have four independent starting values that will affect whether or not a point can escape.

We'll call the "real" and "imaginary" parts of z's starting value as "**A**" and "**B**", and the "real" and "imaginary" increments as "**C**" and "**D**".

The Mandelbrot Set

Starting z from an initial value of *zero*, a plot of the effect of trying different values for the two parts of c, (**C** & **D**) gives us the familiar two-dimensional Mandelbrot Set image.

The Julia Set

For the traditional Julia Set images, we don't start from zero and plot escape as a function of **C** against **D** ... instead, we pick two *fixed* values for **C** and **D**, and plot how changing the initial values of z (**A** & **B**) alter the outcome. **AB** then gives the plotted image, and **CD** gives a way to call up different variations on that image.

If **C** and **D** are both zero, then $z \rightarrow z^2$ gives the central Julia image of a simple non-fractal circle shown in Figure 13-3 (the counterpart of the initial "circle" plot in our hypothetical drinking-glass example). Altering **C** and **D** makes a Julia image morph and swirl – as a generalisation, altering the "real" component **C** makes the image evolve in an orderly way, while varying the "imaginary" component **D** makes the shape break up and twist, rotating clockwise or anticlockwise depending on whether **D** is positive or negative. **D=0** gives us the "core", mirror-symmetrical Julia set images, while setting **C=0** leaves us with just the "twistiness" parameter **D** to play with.

Putting everything together

To summarise, the full Julia Set is four-dimensional, with four freely-adjustable parameters, and the Mandelbrot Set is a two-dimensional slice through it.

The Julia Set was actually discovered first (by **Gaston Julia**), and is usually sliced parallel to **AB** to generate images. Since these shapes seemed to vary in a semi-orderly way when we varied **C** and **D**, Benoît Mandelbrot decided to plot how the images varied with **C** and **D**, and by plotting **CD**, got what we now know as the Mandelbrot Set.

Shadow of the Mandelbrot

Figure 13-1: Crude Mandelbrot map

The image above arranges a selection of standard Julia Set images into a table to highlight any trends (the "Julia Set" equivalent of the Periodic Table). We can see the Mandelbrot Set outline beginning to emerge. The table has parameter **D** increasing from left to right, and **C** from bottom to top. **C=D=0** is marked with a cross.

Since each small Julia image plots **A&B**, and the Mandelbrot Set plots **C&D**, if we deleted everything but the central "0,0" pixel of each Julia image, our table would turn into a pure Mandelbrot Set image.

13. The Julia Set in 4D

Next, here's a higher-resolution table of several thousand Julia images:

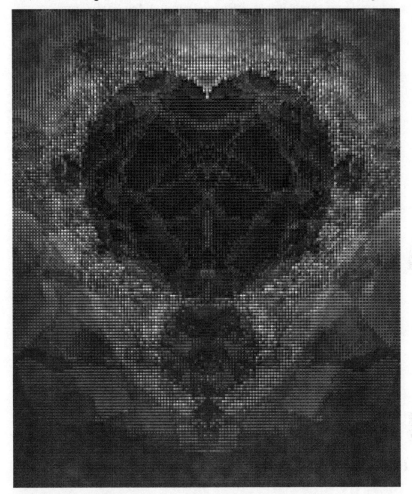

Figure 13-2: A higher-resolution Mandelbrot map

The Mandelbrot Set is sometimes treated as a guide-map to the Julia Set's characteristics. Julia images that correspond to tendril-like locations on the Mandelbrot Set are similarly thin and spiny, side-lobes of the Mandelbrot correspond to Julia images that have left-and-right-handed twists, and as we move towards the middle of a Mandelbrot lobe, the corresponding Julia images get fatter.

As well as the obvious Mandelbrot "shadow", the image shows additional detail from outside the CD plane. It's a projection of four-dimensional detail onto a two-dimensional sheet of paper.

81

So what does the 4D shape look like? Here the six primary sections of our four-dimensional fractal, all sliced through the 0,0,0,0 origin

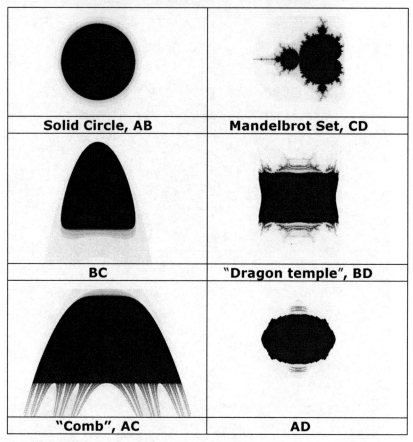

Figure 13-3: Julia Slices

Figure 13-4: "Dragon Temple" detail

13. The Julia Set in 4D

Combinations of our six slices can intersect together to give different three-dimensional "shadows" or partial views of the larger four-dimensional Julia set.

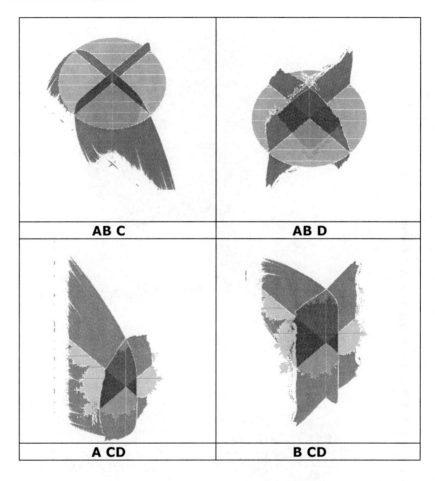

Each of these four frames uses a different three of the four Julia parameters to produce a partial 3D view of the full 4D shape.

We can also plot full three-dimensional images of these solid "shadows".

Solid **AB-C** is a stack of Julia images taken along the central spine of the Mandelbrot Set.

Its mid-region (around the circular cross-section **AB**) has least detail.

There are at least three cross-sections through the shape that give Mandelbrot-like outlines, but without the Mandelbrot's characteristic side-lobe detail – that's contained in parameter **D**.

Figure 13-5: Julia Solid AB C

13. The Julia Set in 4D

 Solid **AB-D** represents a stack of Julia images taken at right angles to the Mandelbrot spine.

Parameter **D** contains the "twistiness" of the Mandelbrot set, and the upper and lower parts of **ABD** twist equivalently in different directions away from the circular outline for **D**=0.

Figure 13-6: Julia Solid AB D

A-CD and **B-CD** both have perfect central Mandelbrot cross-sections.

Figure 13-7: Julia Solid A CD

13. The Julia Set in 4D

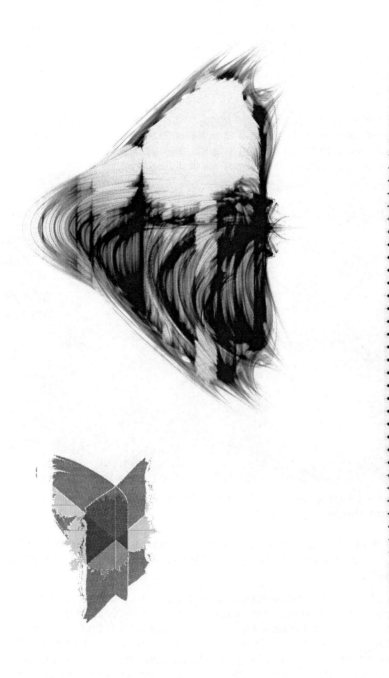

Figure 13-8: Julia Solid B CD

Alt.Fractals

The Julia set contains local styles and themes that evolve across its surface. As a cross-section through the 4D Julia Set, different parts of the Mandelbrot reflect some of the same themes ... but they also contain repeating "variations on a theme" of the complete Mandelbrot image.

The Set contains smaller spawned versions of itself, surrounded by complex islands of detail, and at sufficiently high magnifications, some people have reported finding islands that appear to contain shapes that look suspiciously like full copies of Julia set images.

Figure 13-9: A few "mini-Mandelbrot"-containing regions within the set

This doesn't mean that those Julia Set slices are literally *contained* in the Mandelbrot Set ... it's more a case of what you might call cross-contamination.

The Mandelbrot cross-section is also supposed to have the special property that its entire perimeter is one single continuous line – in theory, we can put a pencil point onto part of the set and return to our starting point after an infinite number of twists, having traced out the entire shape. But since every mini-Mandelbrot must be connected to its parent, it must have external connections that don't exist in the original, and the same applies for any Julia-like regions that we find buried within the Mandelbrot Set.

13. The Julia Set in 4D

Figure 13-10: The Tricorn fractal in Section 15 contains both mini-Tricorns and mini-Mandelbrots

Figure 13-11: Mandelbrot Echoes in solid ACD

A closer look at Figure 13-7 and Figure 13-8 suggests that there may be more than one Mandelbrot set running through the 4D Julia Set.

Since the Mandelbrot Set is normally $z \rightarrow z + c^2$, with an initial value of z = zero, each point on the Mandelbrot Set slice should in theory have a counterpart in the 4D Julia set with an initial value of *minus c^2* that, after one iteration, resets the start value to zero, and then proceeds with a conventional Mandelbrot calculation. So it shouldn't be a huge surprise to find surfaces intersecting the larger 4D shape that seem to produce other Mandelbrot-like cross-sections.

14.   Not the Mandelbrot Set

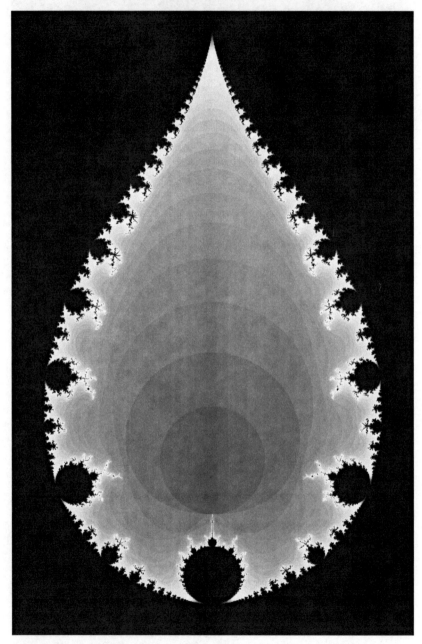

Figure 14-1: Inverse Mandelbrot

14. Not the Mandelbrot Set

Some aspects of the Mandelbrot Set's logical structure are more obvious if we use a mathematical projection to turn the shape inside out around its 0,0 origin.

In the usual plots of the Set, it's tempting to think of the shape of the Mandelbrot as being built from two large components, "a bum and a ball", each with its own branching tendrils and lobes.

The "inverted" version changes the shape's emphasis – the "bum" shape turns into a teardrop, and the left-hand "ball" turns out to be just the largest of a continuous series of bulbs arranged around the teardrop's edge. This "teardrop" version makes it easier to appreciate thematic changes along the row of bulbs: for instance, if we compare the largest sequence of bulbs and count the number of branches at their tips, we find that they form a simple series, 1, 2, 3, 4, 5 …

Figure 14-2: Bubblebrot

15. The Tricorn, or "Mandelbar"

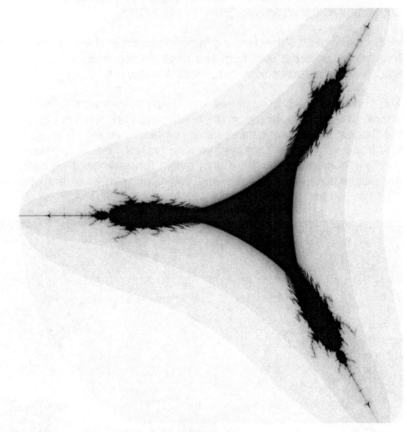

Figure 15-1: The "Tricorn" or "Mandelbar" Set

The complex numbers used to plot the Mandelbrot set have "real" and "imaginary" components that play off against each other. But just as the square root of "plus one" can be "plus one" or "minus one", the square root of *i* can be "plus *i*" or "minus *i*".

Throwing a minus sign in front of the "real" component during the calculations would just flip the shape from left-to-right, but putting a minus sign in front of the "imaginary" component gives a different calculation and instead generates the "**Tricorn**" or "**Mandelbar**" fractal, which has an odd combination of fractal and smooth regions.

Technically, this is referred to as the "complex conjugate" of the normal Mandelbrot, and again, it's self-similar. We can zoom in and find islands that look like copies of the full shape.

15. The Tricorn, or "Mandelbar"

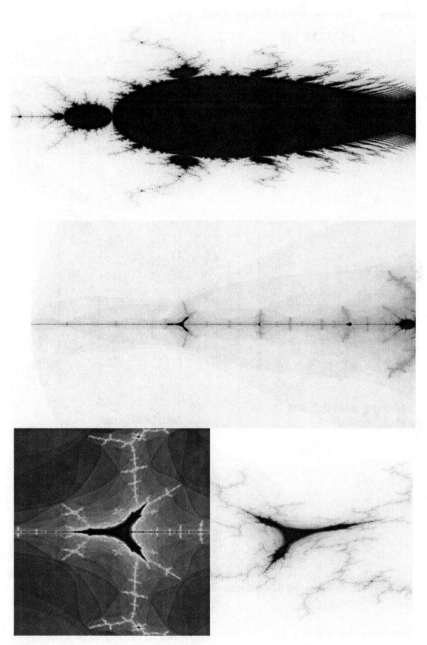

Figure 15-2: Tricorn detail (see also Figure 13-10)

16. The Mandelbrot/Tricorn Hybrid

The Mandelbrot and its "evil twin" share the same "real" axis, which means that they can intersect neatly on two perpendicular planes – we can plot x, y and $\bar{y}$ (known as "y bar") to get a solid shape with both the Mandelbrot and "Mandelbar" shapes as cross-sections

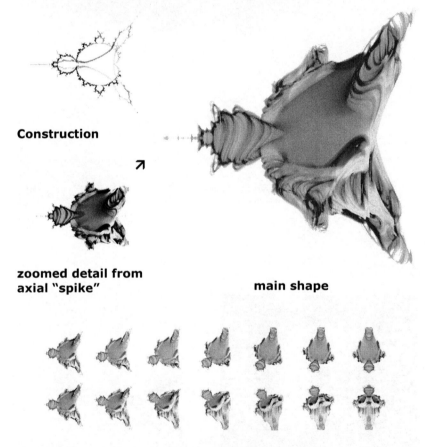

Construction

↗

zoomed detail from axial "spike"

main shape

Figure 16-1: Hybrid Mandelbrot/Tricorn solid

This is more than a simple extrusion. We can see that orbits and structures that circle and move between both planes, notably in the shape's side-bulbs.

16. The Mandelbrot/Tricorn Hybrid

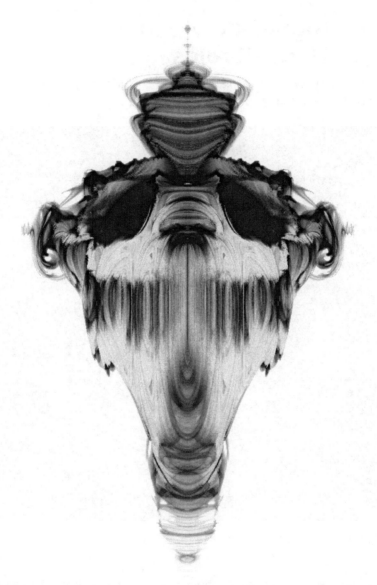

Figure 16-2: Hybrid fractal solid, rear view and zoomed detail

There are also locations where "mini-Mandelbrots" coincide with "mini-Mandelbars", creating miniature versions of the main solid, making it officially fractal.

17. Higher-Powered Mandelbroids

Another way to produce variations on the Mandelbrot image is to swap the z^2 term for z^3, z^4, or another power.

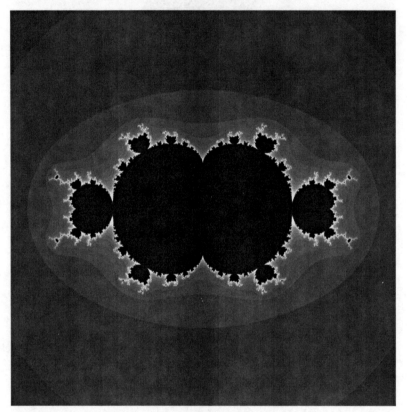

Figure 17-1: Mandelbrot equivalent for z^3

The general rule for these shapes is that the degree of rotational symmetry is one less than the power of z, so z^6 gives a shape with fivefold symmetry, z^3 gives twofold symmetry, and so on.

Each of these shapes also has an associated family of Julia set images.

17. Higher-Powered Mandelbroids

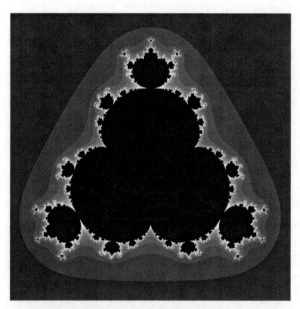

Figure 17-2: Mandelbrot equivalent for z^4

Figure 17-3: ... for z^5

Figure 17-4: ... for z^6

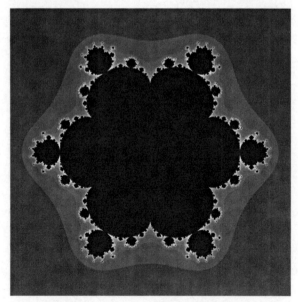

Figure 17-5: ... for z^7

18. Higher-Power Julia Set Maps

Higher-powered versions of the Julia Set can be calculated in the same way.

Figure 18-1: Map of z^2 Julia sets, 25:1 step resolution

Here's the conventional Julia set map, at a scale of one image every 0.04 units. If we locate the central circular Julia image, each step vertically or horizontally represents a change in **C** or **D** of 1/25.

Figure 18-2: z^3 Julia images

18. Higher-Power Julia Set Maps

Figure 18-3: Map of z^3 Julia sets

Replacing z^2 with z^3 gives a different table, containing "alternative" Julia Set images with threefold symmetry. The shadow of this table gives the double-headed z^3 Mandelbrot shape shown earlier, in Figure 17-1.

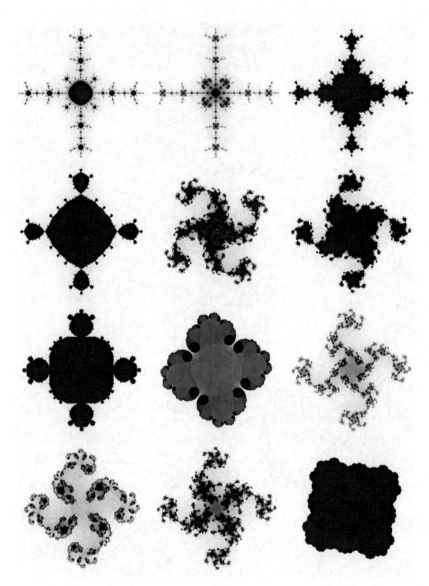

Figure 18-4: z^4 Julia images

18. Higher-Power Julia Set Maps

Figure 18-5: Map of z^4 Julia sets

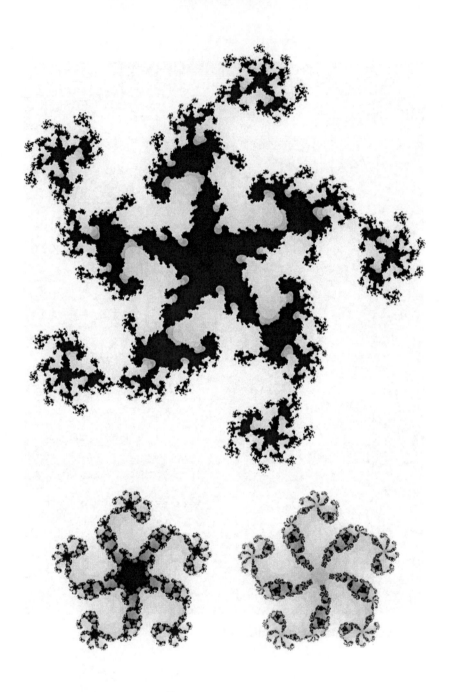

Figure 18-6: Sample z^5 Julia images

18. Higher-Power Julia Set Maps

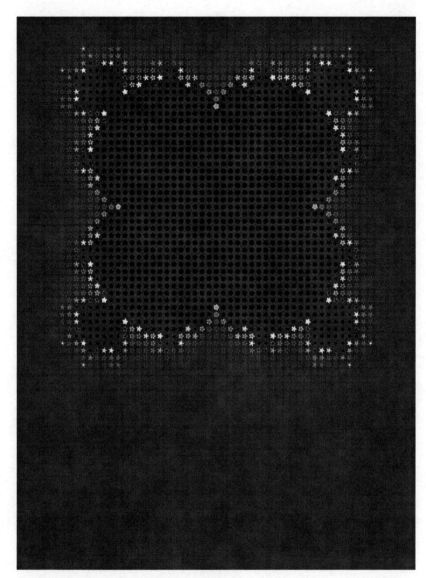

Figure 18-7: Map of z^5 Julia sets

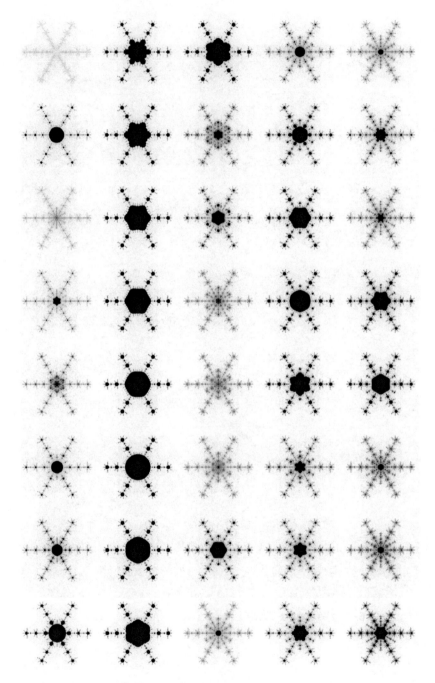

Figure 18-8: A z^6 "snowflake" sequence

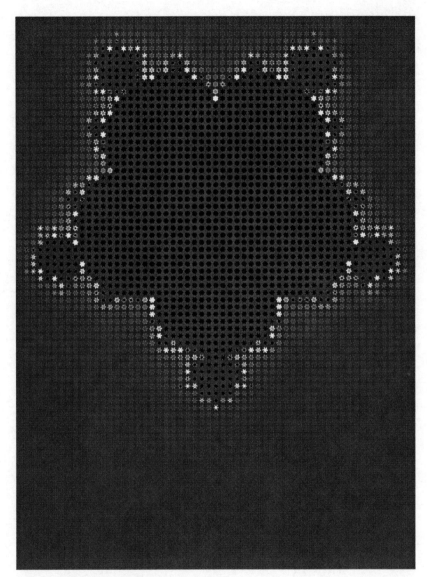

Figure 18-9: Map of z^6 Julia sets

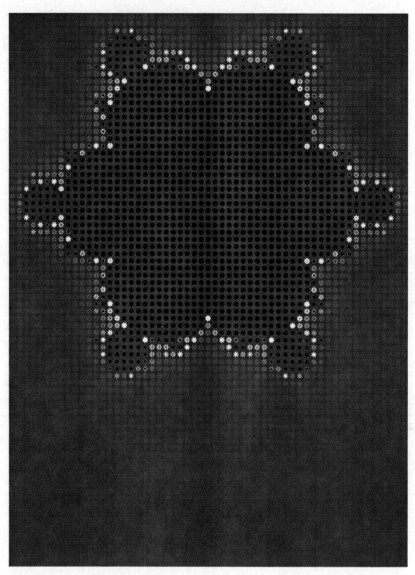

Figure 18-10: Map of z^7 Julia sets

19. More "Complex Conjugate" Mandelbrots

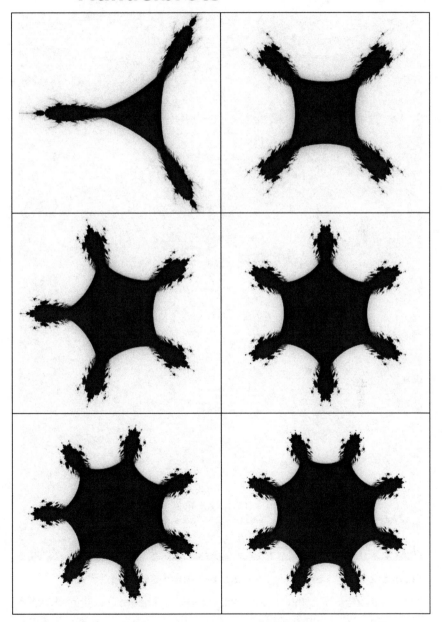

Figure 19-1: "Complex Conjugate" Mandelbrots, z^2 to z^7

20. "Complex Conjugate" Julia Maps

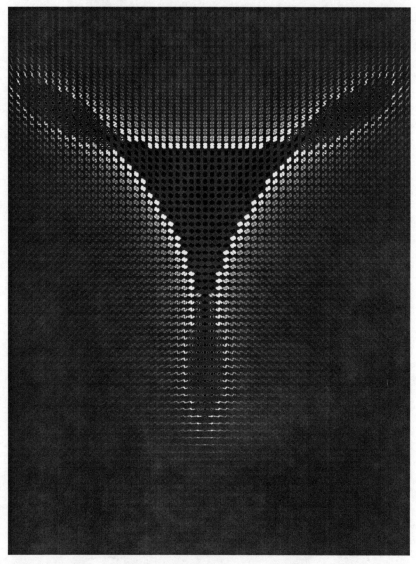

Figure 20-1: "Complex Conjugate" Julia Set Map, z^2

The method of "flipping" a Mandelbrot into a Tricorn (section 15) also works with Julia sets. Here's a table of "flipped" Julia set images, giving a "Tricorn" profile.

20. "Complex Conjugate" Julia Maps

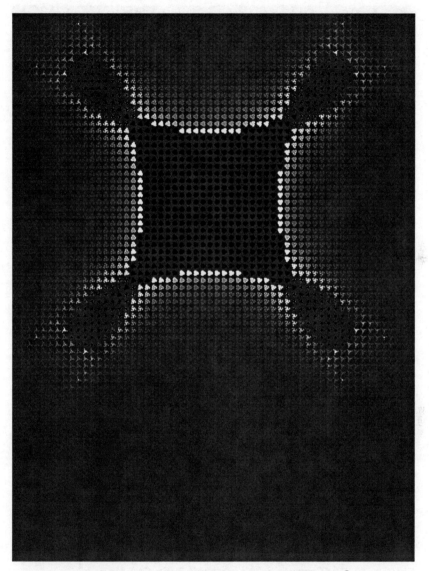

Figure 20-2: "Complex Conjugate" Julia Set Map, z^3

Higher-powered Julia set calculations can also be "flipped" in the same way. Here are tables of the "alternative" Julia set images, from z^3 to z^5. Their outlines are the "higher-powered complex conjugate Mandelbrot" shapes from Figure 19-1.

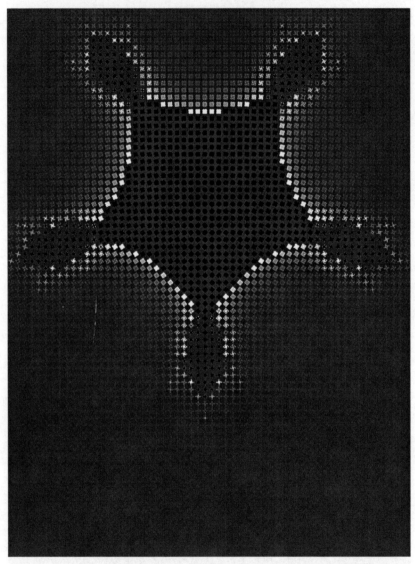

Figure 20-3: "Complex Conjugate" Julia Set Map, z^4

20. "Complex Conjugate" Julia Maps

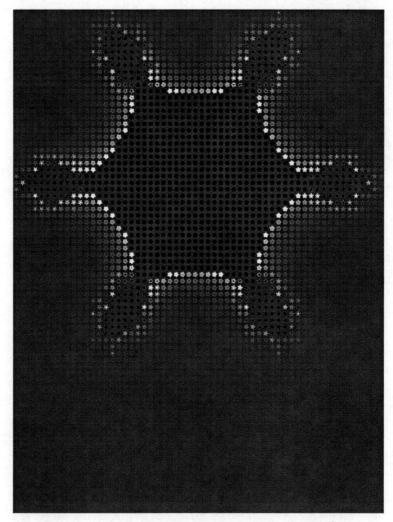

Figure 20-4: "Complex Conjugate" Julia Set Map, z^5

21. Higher-powered Julia Solids

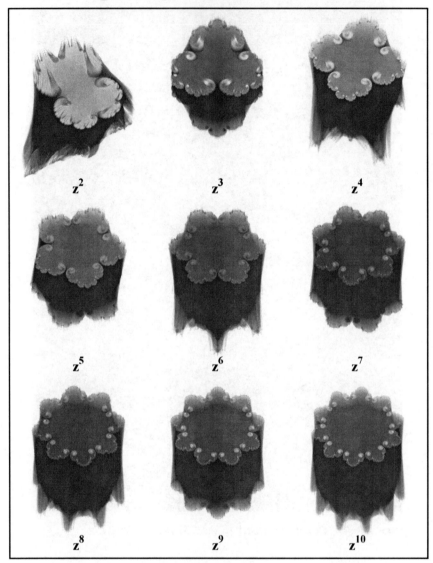

Figure 21-1: Julia Solids: ABC_ (core set, powers 2-10)

Each of the higher-power groups of Julia sets also has its own 4D solids, as before (although these can involve a certain amount of repetition). The "**ABC_**" solids correspond to a "stack" of Julia images taken from the central column of each map (with **D=0**).

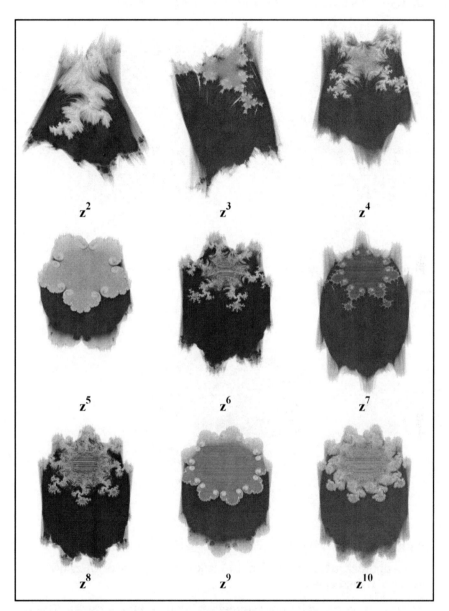

Figure 21-2: Julia Solids: AB_D (powers 2-10)

The **AB_D** solids correspond to slices *across* the Mandelbrot map axis. They have a "twist" whose strength and polarity depends on **D**.

22. Higher-power Hybrids

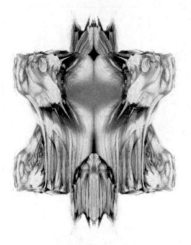

Figure 22-1: z^3 hybrid

Figure 22-2: z^5 Hybrid

The approach used in section 16 works for any power of z, giving a family of hybrid solids.

Each version has different symmetries around different axes. The z^2 solid (Figure 16-2) has symmetries of one and three, the z^3 has symmetries of two and four, the z^4 has three and five, the z^5 has four and six, and so on.

22. Higher-power Hybrids

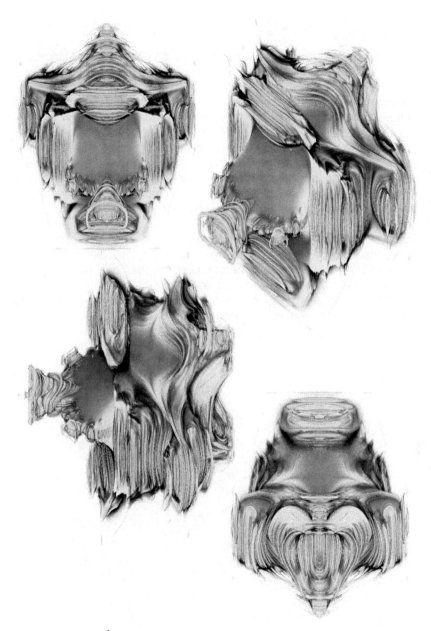

Figure 22-3: z^4 Hybrid

23. Hyper-Complex Fractals

Another way to create three-dimensional shapes from a Mandelbrot Set is to add a second independent "imaginary" value, at right angles.

Putting conventional imaginary terms on the y- and z-axes gives a simple solid that represents the Mandelbrot spun around on its axis, but this isn't very interesting. For the Mandelbrot/Tricorn hybrids, we kept the two imaginary terms distinct by giving them different signs: this gave a "proper" complex shape, turning the ox plane from a Mandelbrot into a three-cornered "Tricorn". To force *two* intersecting Mandelbrots while making sure that something interesting happens at intermediate angles, we can use **hypercomplex numbers**.

Where *complex* numbers have a one "real" and one "imaginary" component, *hyper*complex numbers can have multiple independent imaginary parts. A simple hypercomplex number has two imaginary terms (i and j), representing different square roots of minus one. Multiplying i times j gives a fourth creature, k, and k-squared gives *plus* one. A bit more math can then give us Figure 23-1.

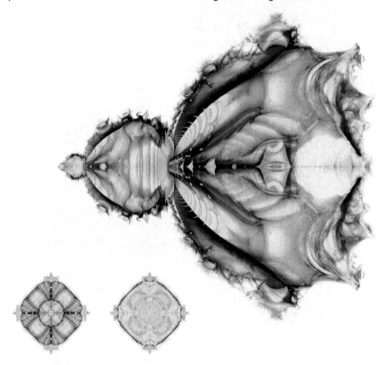

Figure 23-1: A Hypercomplex Mandelbrot Set

23. Hyper-Complex Fractals

With our "real" component r and our two separate, equivalent (but different) imaginary terms i and j, we can go ahead and plot a three-dimensional shape knowing that it'll give conventional Mandelbrots on two axes, but that because of the complicating k term, the shape at intermediate angles will be something more complex than you could produce on a lathe.

We can also have hypercomplex multi-dimensional Julia sets (with some duplication), and since there are so many different ways of constructing number systems with multiple imaginary components (for instance, **quaternions** and **octonions**), there are many ways of tweaking these solids to get different results, by carefully selecting different number systems.

The downside of this is that the choice of shape for these hypercomplex fractals can sometimes seem a bit arbitrary.

Figure 23-2: A Hypercomplex Mandelbrot Set for z^4

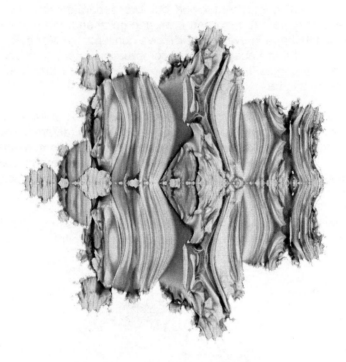

Figure 23-3: A Hypercomplex "Mandelbulb", z^8

Hypercomplex versions of the z^8 Mandelbrot (known as the "**Mandelbulb**") can have a particularly nice lobed structure, but didn't seem to be properly explored and modelled until Paul Nylander and Daniel White attacked them in 2009.

24. Sine Julia Sets

Replacing the Mandelbrot/Julia z^2 term with some other function of z gives even more types of fractal.

Using Sin (z) produces extended, ribbon-like fractal shapes, which can sometimes look a little bit like of wallpaper edging designs.

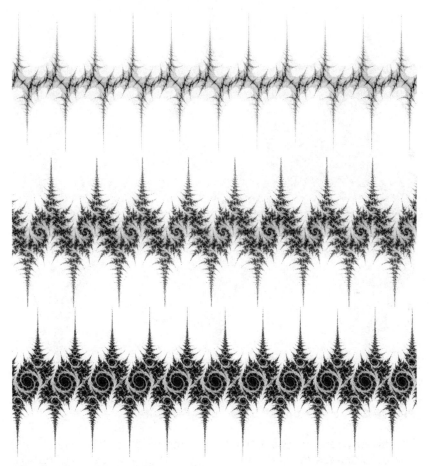

Figure 24-1: Sine Julia Images

Since they can include contributions from an infinite number of positions along the x-axis, some of these "Sine" Julia plots can have be extremely dense internal structure, which can make them difficult to plot.

Alt.Fractals

To convert these into useful images we can create "compacted" versions where we plot a secondary, derivative function based on the amount of detail at each location, or we can simplify the calculations by adding a fifth parameter (a "bailout" value) that halts the calculation if the coordinates move too far out of range.

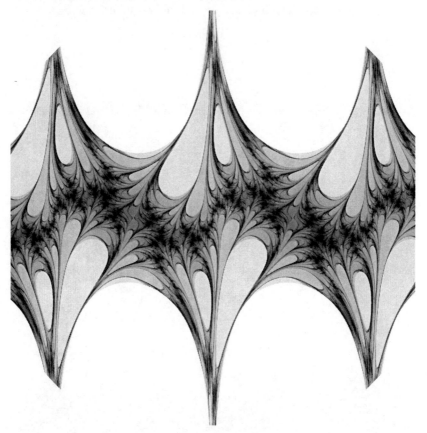

Figure 24-2: "Stripped" Sine Julia Fractal

We already tend to use bailout functions when plotting Mandelbrot and Julia images to speed up the calculations – we assume that if a point has moved as far away as, say, four units, it's already irretrievably lost and we don't need to do any further calculations on it.

This approach can change internal detail for some "trigonometric" fractals, because where the fractals form infinitely long strips, a point can be infinitely far from its starting point (along the x axis), and still be trapped.

24. Sine Julia Sets

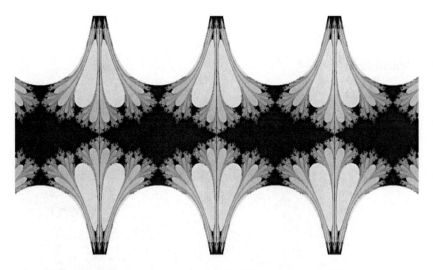

Some features in trigonometric fractals look rather similar to features in standard Julia sets or the Mandelbrot set. Thanks to **Pythagoras' Theorem** ("*The square of the hypotenuse is equal to the sum of the squares of the other two sides*"), some trig functions tend to have internal z^2 relationships.

Figure 24-3: Sine fractal detail

"Standard" & "Sine Julia" images

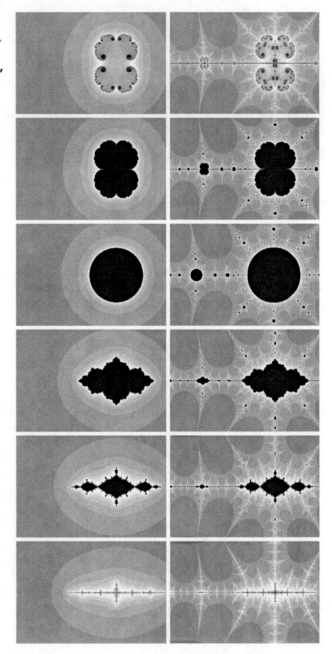

Figure 24-4: Some "core" Julia images and their "Sine Julia" counterparts

24. Sine Julia Sets

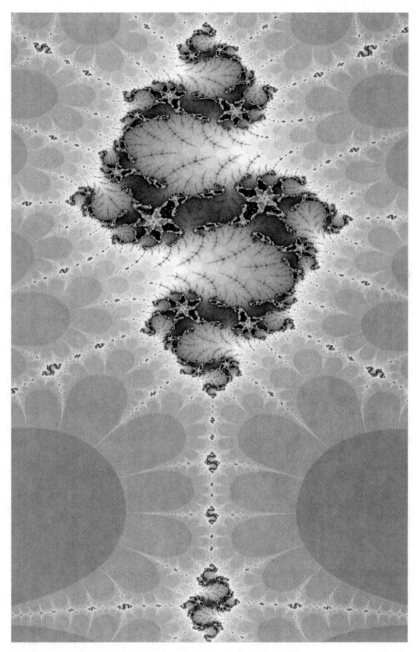

Figure 24-5: Repetition in a Sine Julia fractal

Figure 24-6: Another partial Sine Julia image

25. Other Trigonometric Julia Sets

We're not forced to limit ourselves to only trying sine functions. Trigonometry's "tangent" function (which relates the angle of a right-angled triangle to the relative lengths of the two perpendicular sides) also creates some interesting shapes when we feed it into the iterative sausage-machine, and there are also other, more exotic functions to play with.

Figure 25-1: "Rope" Fractal using a TAN function

The "tangent" function tends to generate shapes with a cross-like character, and again, an additional "bailout" parameter can be useful for limiting what might otherwise be overwhelming detail.

Figure 25-2: A tan-function fractal: Worlds within worlds

25. Other Trigonometric Julia Sets

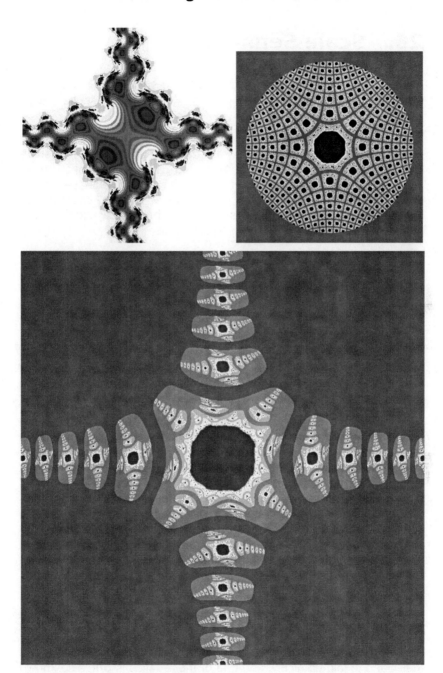

Figure 25-3: More trig-function fractals

26. Scale Sensitivity

Although we often describe fractals as being similar at all scales, the Mandelbrot set ... which to many people is the quintessential fractal ... isn't.

We can zoom *in* on the Mandelbrot Set infinitely far and always see further detail, but if we zoom *out*, the detail stops coming. The entire Mandelbrot Set has a limited extent, and since it's reckoned to be singly-connected, once you've included the entire perimeter of the shape, it's guaranteed that there's nothing else outside. Zoom out on the Mandelbrot Set and it shrinks to a blob, then a dot, and then to a dimensionless point. There's nothing else to see.

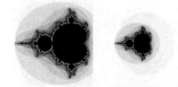

On the other hand, the Sierpinski Triangle lets us zoom in *or* out indefinitely, and keep adding siblings alongside the largest chunk of fractal in view to make larger copies. These copies are "similar" in the mathematical sense, but in more everyday language they're not just similar at all scales but *identical*.

The Mandelbrot's "dead end" limit arguably gives its patterns a reference and sense of scale that's lacking in the Sierpinski Triangle. The Mandelbrot Island only appears in its exact form, with no external connections, at one particular scale, and the degree of potential complexity increases as we zoom in. Particular types of shape can only be found below a certain depth, and so find *some*, you have to zoom in further than 16-bit floating point calculations will allow. The scaling-distance from the top-level shape gives a sense of size to at least some of the fractal's features that's missing from some other fractals that really *are* similar at all scales.

So are there any fractals that turn this property inside out? Do any fractals allow us to zoom *out* infinitely far, but come to an end when we zoom *in*?

Yes. Some fractal types have to be drawn with particular ratios to get their parts to join up nicely, and one of these ratios is **phi**, also known as the **Golden Section** (there are some "*phi*" fractals in sections 27 to 29). *Phi*-proportioned fractals aren't particularly novel, but since the pairs of numbers in the **Fibonacci Series** have ratios that *converge*

26. Scale Sensitivity

infinitely closely on *phi*, we can take a "phi fractal" and replace its proportions with those of consecutive pairs of high-sequence Fibonacci numbers without seeing any apparent difference. But if we zoom in on these fractals, eventually the Fibonacci sequence will run out as the series counts down to one. At the end of the range, the proportions start to veer away from *phi*, and then the pattern abruptly stops. It has a smallest possible building block, and is what you might call an "atomistic" or "quantum" fractal.

Figure 26-1: Interlocking Face-Corner Fractal Tiling

Consider the pattern in Figure 26-1. It's a deterministic tiling that we can extend outwards infinitely far, and at all iterations we have four triangular blocks, all butted up against each other. At larger scales the blocks appear significantly interleaved, and the shape converges on something that looks like the shadow of the Sierpinski Pyramid, with three corner triangles that are vanishingly close to touching, and a fourth, overlaid (in this case enmeshed) copy in the centre. The proportions don't seem to change perceptibly as we zoom further out, but if we zoom *in*, the corner-pieces move proportionally further apart until we reach maximum zoom, and the pattern stops. Once we reach the shape's "atomistic" scale, we can't subdivide the shape further. If we try to divide the fractal's "atom", the fractal's pieces no longer touch, and if we adjust their positions to correct for this, the entire fractal has to be redrawn.

Figure 26-2: Interlocking Face-Corner fractal tiling (alternative tiling mode)

26. Scale Sensitivity

Figure 26-3: Interlocking Face-Corner Tiling (alternating modes)

Figure 26-4: ... Tiling the Plane

Alt.Fractals

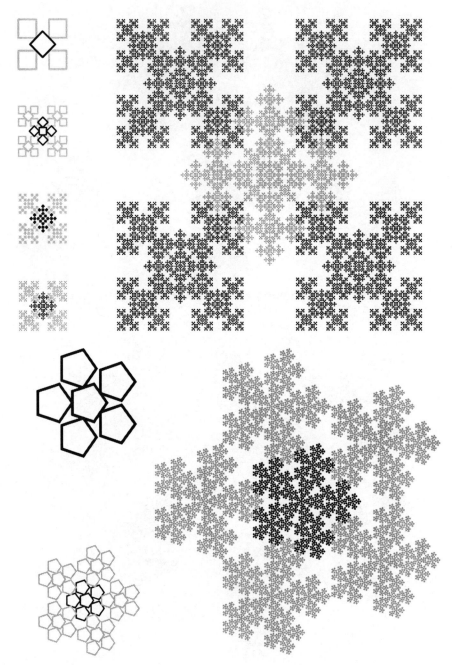

Figure 26-5: Corner-Side Fractals, using squares and pentagons

26. Scale Sensitivity

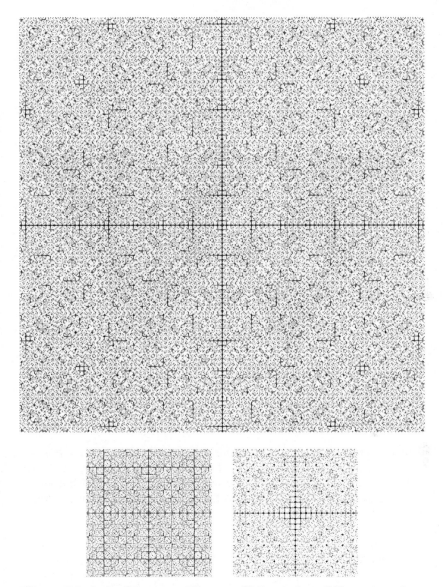

Figure 26-6: An "Inverse Fibonacci" fractal at different scales

Figure 26-6 runs the Fibonacci method backwards, using two start values representing positions on the *x* and *y* axes. Normally, the method makes adjacent pairs of numbers converge on *phi*. Used in reverse, it produces increasingly wild swings. Pixel colouration is based on the nature of the "crash" that happens if and when values drop below one. The shape's characteristics vary with magnification.

27. Fibonacci and the Golden Section

Some ratios have a special significance in fractals and in branching systems in general. The first of these is the "binary series" ("Base Two"), where we double or halve our quantities at each stage. The next is Base Three, where we use ratios of one third.

Figure 27-1: Fibonacci Rose

After this, we have the **Golden Ratio ("phi", ~1.618034...)**, and by association, the **Fibonacci Series**.

The **Fibonacci Rose** neatly illustrates the principle of the Fibonacci Series: each triangle has a face that butts against the faces of two smaller triangles representing the two previous sizes, and the length of each triangle's face is equal to those of the previous two sizes of triangle added together.

If the two smallest black or white triangles have length "1", then the sequence of sizes, for either black or white, goes

1, 1, 2, 3, 5, 8, 13, 21, 34, 55, 89, 144 ...

This is the standard Fibonacci series.

We can also consider the Fibonacci series as describing as series of branchings, with the **extended Fibonacci Series** (section 29) allowing a variation in how the branching happens.

27. Fibonacci and the Golden Section

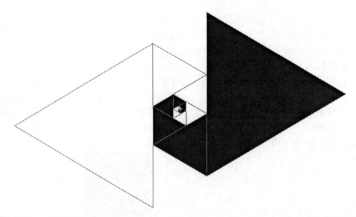

Figure 27-2: "Fibonacci Rose" grid, with alternative shading

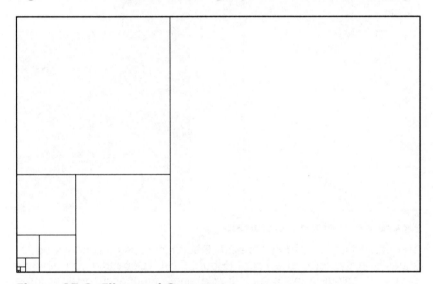

Figure 27-3: Fibonacci Squares

Fibonacci Tilings usually use squares rather than triangles. Figure 27-3, starts with two unit squares butted together to make a rectangle, and adds a new square to the rectangle's longest side. The new square's size is "2×2", and in conjunction with the two earlier squares, makes a 2×3 rectangle. Adding a 3×3 square to *its* longest side then gives a rectangle of size 3×5, and the longest sides then increase according to the Fibonacci Series as before: 8, 13, 21, 34, and so on.

The rectangle alternates between "thin" and "fat" outlines, converging on a critical central proportion known as the **Golden Ratio** or **Golden Section**, which us sometimes assigned the Greek letter "*phi*" ("ϕ").

Alt.Fractals

Like "pi" ("π"), the ratio of the circumference of a circle to its radius, "phi" is an "irrational number" that can't be expressed using a finite number of digits, but the value comes out as approximately **1: ~1.6180...** . Inverting the ratio, (1/phi), gives the same sequence of digits with the initial "one" replaced by a zero, (~0.618...) Tilings using the Fibonacci series of integers can be replaced with tilings using the Golden Ratio and still produce a perfect mesh, although it'll now have no lower limit to the tiling detail, and its scaling will extend upwards *and downwards* infinitely far.

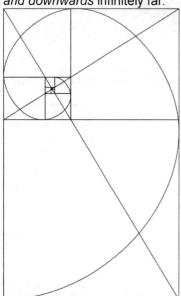 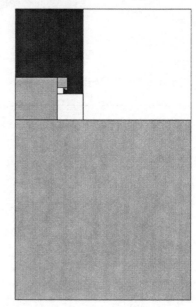

Figure 27-4: Fibonacci Spirals

The other standard "square-based" Fibonacci construction lays out its squares in the form of a spiral (Figure 27-4). Depending on how we choose to colour it, this gives either a single Fibonacci spiral, or three interlocking "Fibonacci-cubed" spirals

The ratio of "one half" appears regularly in geometrical fractal designs in this book, because the series

$$1/2 + 1/4 + 1/8 + 1/16 + 1/32 + 1/64 + 1/128 \dots$$

converges neatly on a value of "one". Similarly, "thirds" also appear regularly because the corresponding series

$$1/3 + 1/9 + 1/27 + 1/81 \dots$$

converges neatly on "one half".

27. Fibonacci and the Golden Section

With a ratio of 1:2, adding an infinite chain of smaller copies to a shape, end-to end, gives a chain that will have exactly the same overall length as the unit shape that spawned it. With a ratio of 1:3, we can tile from both ends of the parent shape, and both infinite chains should meet perfectly in the middle. Both series are regularly used to create fractal patterns whose parts fit together perfectly.

Phi doesn't appear quite as often as 1/3 and 1/2, but it shows up whenever we involve regular pentagons (section 30), and it also appears in a range of "successive approximation" tiling systems involving other shapes.

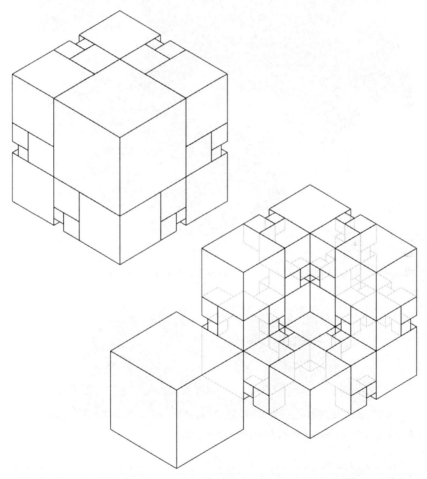

Figure 27-5: Fibonacci /Golden Section stacking of 3D blocks

Figure 27-6: Fibonacci / Golden Section "Oak Tree"

Figure 27-7: Fibonacci-version detail

Returning to the subject of fractals, if we take one arm from the double-spiral in Figure 27-2 and add a new arm to every unused face, we get the fractal in Figure 27-6, which (thanks to the Golden Section) has a series of corners that *just* touch.

27. Fibonacci and the Golden Section

Figure 27-8: Golden Ratio Triangles

Another fractal shape that needs the Golden Ratio in its construction is the "corner-triangles" shape in Figure 27-8. This was the basis of the 3D "corner-tetrahedra" shape in Figure 8-5, which also used the Golden Ratio.

From the point of view of the "atomistic" fractals in section 26, it's significant that we can often turn "Golden Section" fractals into their "atomistic" counterparts by replacing the ratio *phi* with the ratio between adjacent high-value pairs in the Fibonacci series. When we're zoomed out a long way, the two variants will be indistinguishable, and their parts will exactly interlock in the same way regardless of whether they're scaled using "phi" or Fibonacci. But if we're looking at the "Fibonacci" variant, and we zoom in far enough, the ratios will eventually swerve away from the ideal and "crash", so that the sequence ends with a final building-block that can't be divided any further.

28. Tiling the Golden Rectangle

If we start with a Golden Ratio rectangle, cut it corner-to-corner, and try to tile the resulting triangle with maximally-sized squares, something odd happens. The sizes of the row of squares running along the bottom of the figure have a Golden Ratio relationship, and the column of squares on the left hand side have sizes locked to *phi*-squared. Every tile in the pattern is part of a single Golden-Ratio sequence, and the triangle is completely tiled using just the (infinite series of) sizes of square found along the lower row.

The resulting interlocking pattern is almost crystalline. If we allow the parent rectangle the tiniest deviation in proportions away from the Golden Section, the system breaks down – we can still tile, but the squares no longer fit into a simple series.

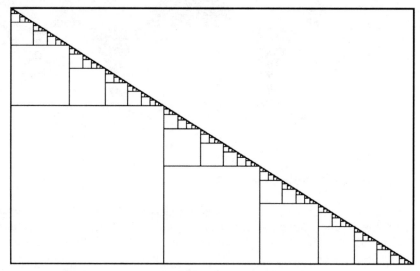

Figure 28-1: Fibonacci-tiling a Golden Rectangle

Because all the squares fit into a single Golden Ratio sequence, we can easily rank and group them by size. We have one large square in the bottom left hand corner, one smaller square to its immediate right, and then two squares the same size (square #3 on the bottom row, and the square directly above the corner-square).

The next size down has three copies, the one after that has five. Writing out more of the "square-counting" sequence gives

1, 1, 2, 3, 5, 8, 13, 21, 34, 55, 89, 144, …

… We're back to the Fibonacci Series again.

29.   Extended Fibonacci Series

The Fibonacci series and Golden Ratio aren't the only ways to get this sort of quantised tiling. If we smoothly vary the proportions of the initial rectangle to make it fatter, there turns out to be only one other solution that allows the squares to snap into simple ratios, and that's the one that appears when the smallest angle of the triangle is 45 degrees.

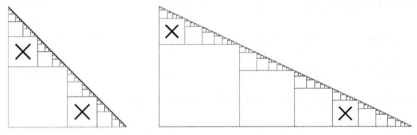

Figure 29-1: Tilings for *eF*(0) and *eF*(2)

"Counting the squares" then gives quantities of:

1, 2, 4, 8, 16, 32, 64, 128, ...

It's simply "powers of two".

If we try to make the rectangle thinner than the Golden Ratio, the next quantised solution shows up at around ~25 degrees, and gives the sequence:

1, 1, 1, 2, 3, 4, 6, 9, 13, 19, ...

, and the one after that gives:

1, 1, 1, 1, 2, 3, 4, 5, 7, 10, ...

After this, there's a sequence that starts with five consecutive "ones", then six, then with seven ... logically, there have to be an infinite number of these quantised tiling sequences (with increasingly-narrow associated angles), and on investigation, all of these sequences turn out be members of something called the **extended Fibonacci series**.

Alt.Fractals

With the normal Fibonacci series, each number is the total of the number on its immediate left, and the number before that. With the "extended" series, the positional offset of the leftmost number varies.

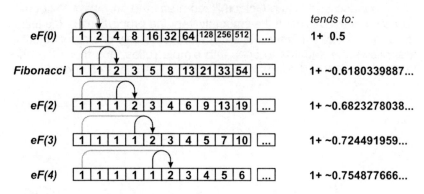

tends to:

eF(0) `1 2 4 8 16 32 64 128 256 512 [...]`	1+ 0.5
Fibonacci `1 1 2 3 5 8 13 21 33 54 [...]`	1+ ~0.6180339887...
eF(2) `1 1 1 2 3 4 6 9 13 19 [...]`	1+ ~0.6823278038...
eF(3) `1 1 1 1 2 3 4 5 7 10 [...]`	1+ ~0.724491959...
eF(4) `1 1 1 1 1 2 3 4 5 6 [...]`	1+ ~0.754877666...

Figure 29-2: Extended Fibonacci sequences

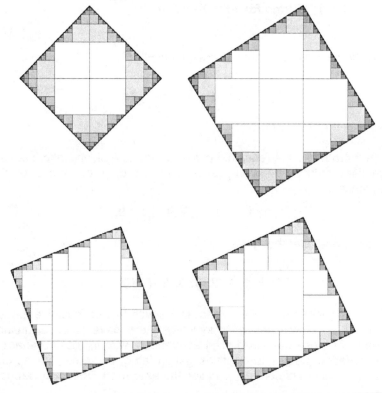

Figure 29-3: Extended Fibonacci Tilings (0 to 3)

29. Extended Fibonacci Series

Figure 29-4: "Extended Fibonacci" *eF*(**3**) **Spiral**

A familiarity with the extended Fibonacci family can be useful in fractal tiling problems. If we want to construct Figure 29-4, we might happen to notice that height of the two vertically-stacked triangles on the right, added together, equals the height of the largest triangle. If the ratio between each triangle and its immediate offspring is **1:r**, then the ratios between the central triangle and these two successive offspring will be **1:r^2** and **1:r^3**, so we know that the value of **r** that we need to be able to draw this diagram has to satisfy the rule **r^2 + r^3 = 1**. Some calculator-punching then tells us that **r** should be somewhere around ~0.754878... . But if we know about the extended Fibonacci series, it's simpler to recognise that this diagram represents a tiling based on extendedFibonacci(4), and simply look up the ratio's value in a table.

The other reason for involving the extended Fibonacci series is that small rounding errors raised to high powers can cause cumulative mismatches in computer graphics. If the correct ratio for a diagram is an irrational number, it can't be expressed correctly by any number of digits in a conventional computer system – it'll always be wrong, and these tiny mismatches can be magnified by rounding-errors and repeated tiling. If we wanted to use the design for an architectural floorplan, we'd want the corners to line up precisely. If we take a sequence of large integer values from a Fibonacci-series sequence and use these as the nominal sizes of our triangles instead of using the "proper" ratios, we can guarantee that even though the proportions may stray imperceptibly from the ideal, the resulting alignment of the tiles won't just be close, but *perfect*.

30. The Golden Ratio and Pentagons

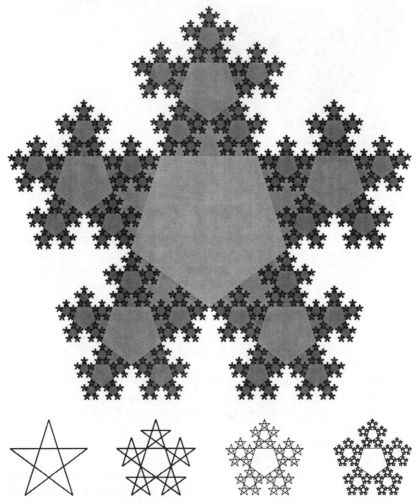

Figure 30-1: Pentangle "Briar"

The **Pentangle Briar** has some similarities with the Koch Snowflake. The ratio between the widest part of a pentagon and the length of its base is *phi* (the Golden Ratio, sections 27-30), so this ratio crops up repeatedly when we deal with fractals having fivefold symmetry. Pentagonal Briars can also interlock and tile with each other, corner-to-face, if their relative sizes fit the Golden Ratio (Figure 30-4).

30. The Golden Ratio and Pentagons

Figure 30-2: Subdivided Filled Pentagon

We can also produce the same outline as Figure 30-1 by using rings of pentagons rather than pentangles – although the initial shapes are different, they converge on the same final result.

Filling the hole in the centre of the ring with a sixth pentagon creates interesting internal structure that looks rather like a slice through an apple-core or plant stem.

Figure 30-3: Pentagonal "Briar Curve" (triangular division method)

Figure 30-4: Pentagon Briar Tiling

The pentagonal "**Briar Curve**" is an inevitable intermediate between Figure 3-18 and Figure 3-19, and tiles in a similar way to the Koch Snowflake. The corner angle of the parent triangle is 36 degrees.

30. The Golden Ratio and Pentagons

Figure 30-5: Pentagon Tiling

Sometimes it's the way that a system fails that creates the most interesting patterns. Figure 30-5 appears to represent the lowest level of another "atomistic" fractal. Starting with a central briar-like "spiked star", the system progressively tiles pentagons, corner-to-corner. The geometrical frustrations inherent in attempted pentagonal tiling repeatedly cause the system to break, at which point, we allow a void, insist that the void has fivefold symmetry, and continue. The system seems to generate a sequence of ever-larger alternating "star" and "flower" voids as we move further from the centre, and since there shouldn't be any pentagonally-symmetrical shape that tiles simply, this escalation presumably continues forever at larger and larger scales, giving a fractal sequence of larger and larger holes.

31. Islands and Double-Limits

We can sometimes impose an artificial upper limit on fractals that would otherwise extend infinitely outwards by "**islanding**": closing the sequence around on itself so that no larger copies are possible. The Koch Curve (Figure 2-6, Figure 34-1) was a good example – in theory, the curve let us zoom in or out infinitely far, but when we connected three Koch Curves together to produce the standard Koch snowflake, the shape "closed". It became a self-contained island, preventing any more detail from appearing when we zoomed out further (although the variants in Figure 7-4 and Figure 7-5, which contain full copies of their island within themselves, don't have this limitation).

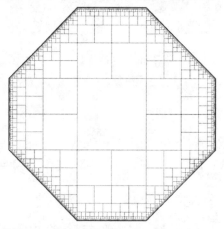

Figure 31-1: Square Cluster

We can use "islanding" to create upper limits for other infinite fractals. Koch snowflakes can tile space perfectly using an infinite range of sizes, but once we create the compound Koch island in Figure 31-2, the fractal tiling gets an upper size-limit (unless we start using a different rule to "tile the tiles").

"Islanding" challenges conventional definitions of what a fractal is supposed to be. We're already familiar with the idea that the Mandelbrot and Julia sets have an upper limit, and are only infinite in one direction – we can zoom in on them as far as we like, but zooming out brings us to a dead stop as far as revealing further detail is concerned. Even though the Mandelbrot and Koch Snowflakes aren't technically self-similar at *all* scales (only at all *smaller* scales), we choose to overlook this because we know that they still have an infinite range.

31. Islands and Double-Limits

The "atomistic" fractals turn the behaviour inside out. They let us zoom *out* as far as we like, but have a limit on how far we're allowed to zoom *in*. Since these still have an infinite range, the fact that this range only applies to zooming *out* shouldn't make them any less legitimate than the Mandelbrot Set and Koch Snowflake.

Figure 31-2: Snowflake-sequence Tiling

The definitional challenge appears when we realise that we can use "islanding" on an atomistic fractal to produce a shape that has both upper *and* lower limits. The three-armed shape in Figure 27-6 is a legal fractal (with an upper limit) if it represents a Golden Ratio series. If we substitute a high-valued Fibonacci series, there might be no discernable difference in the shape, and perhaps we might have to zoom in an astronomical (or cosmological!) distance before we'd have any indication that this wasn't a 100%-legal "*phi*" fractal. But if it *is* a Fibonacci fractal, the number or triangles in the figure will be finite, and theoretically countable. Depending on the Fibonacci number-pairs selected for the top-level triangles, the total count might be less than a hundred, or it might be greater than the number of possible states in the observable universe.

Alt.Fractals

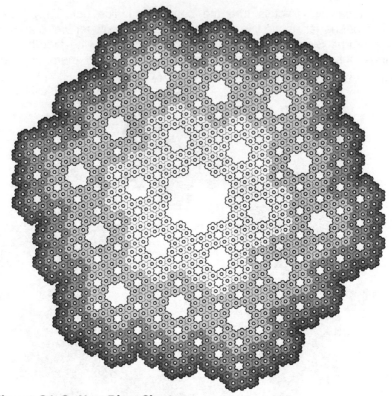

Figure 31-3: Hex Ring Cluster

So do these doubly-finite variations really deserve to be called "fractal", or are they "pseudofractal" phoneys? If part of our motivation for studying fractals is that they appear to mimic features of our own universe, then it's worth pointing out that our observable universe may be "pseudofractal", too. Quantum mechanics gives physics an identifiable lower limit for structural scale that colours the shape of larger structures. A small star has different properties to a large one, a grain of sand is different to a planet, a planet is different to a star, a solar system looks different to a galaxy, and a galaxy supercluster looks different to a cluster. At the largest observable scales, we see new wall-like structures, and above that we have the cosmological horizon preventing us from being able to see whether we inhabit a closed island bubble or a much larger system.

Since we originally avoided studying fractals because they didn't appear to obey the "proper" rules of conventional geometry, it'd be ironic if "fractallists" then avoided studying *these* shapes for the same basic reason, because they didn't appear to fit our current definitions.

32. Linear fractals

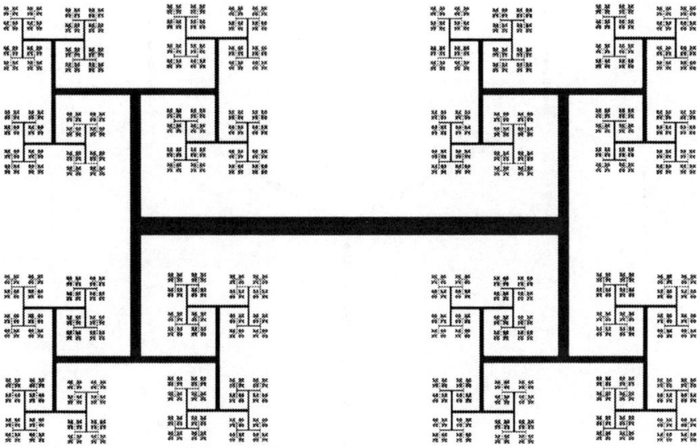

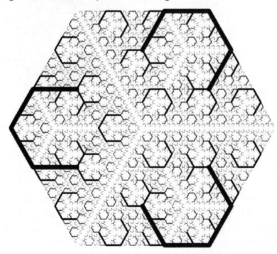

Figure 32-1: "TV Aerial" Recursion

The building-blocks of a fractal can come in various forms. The "filled" Koch Snowflake is often considered as a collection of flat triangles all aligned in a single plane, and the Menger Sponge as an array of 3D blocks in 3D space. But we can also build fractals from lines or planes embedded in three or more dimensions. Complex fractals also tend to generate lower-dimensional fractals as shadows and cross-sections – outlining the profiles of features in the "everted" Menger sponge in Figure 3-10 gave the linear pattern in Figure 32-2.

Figure 32-2: An Everted Menger Sponge edge-profile

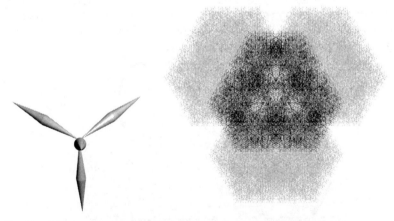

Figure 32-3: 2D Branching ...

Figure 32-4: ... and 3D Tetrahedral-skeleton fluffball

Figure 32-3 and Figure 32-4 can both be considered as fractal skeletons of other fractals: the first can be considered as a skeleton of a triangle-based design with each triangle is replaced by three lines radiating from the centre to its corners, and "skeletonising" the corner-linked tetrahedral fractal from Figure 8-5 then gives a "fluffball" cloud of lines in three dimensions.

32. Linear fractals

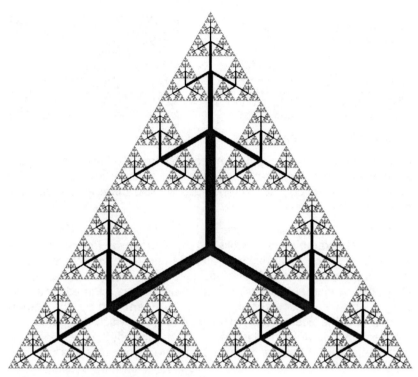

Figure 32-5: Sierpinski network

Figure 32-6: "Single-generation" Sierpinski Triangle and Pyramid network diagrams

The network map of a fractal is usually another (linear) fractal. Figure 32-5 and Figure 32-6 are branching networks representing the parent-child and sibling relationships inside the Sierpinski triangle. Since the map is denser in regions where the original shape has most detail, and has the same basic density-distribution, the maps ends up replicating some of the characteristics of the original shape.

Alt.Fractals

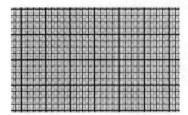 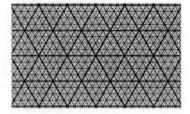

Figure 32-7: Simple Grids

Simple graphpaper-style grids are arguably fractal. They're self-similar at all scales and their patterns also have spacewise repetition. But they're not especially interesting unless we use them as the basis of more complex shapes.

One of the simplest alternatives is the family of space-filling curves whose first example was published by **Giuseppe Peano** in 1890. These fill a space uniformly with a single unbroken non-crossing line, to arbitrarily-high density. There are quite a few different variations on the most fundamental versions of these curves, named after the people who first discovered them (like the **Peano Curve**, the **Hilbert Curve**, and the **Sierpinski Arrowhead Curve**, which navigates only the sections of a triangular grid that are included in a Sierpinski Triangle). **The "H" curve** (Figure 32-9, over) populates the plane by tiling it with "H" outlines, linked in clusters of four by additional "H"-outlines, to give progressively larger H-shapes that rotate by 90° at each generation.

Figure 32-8: Simple and complex space-filling curves

32. Linear fractals

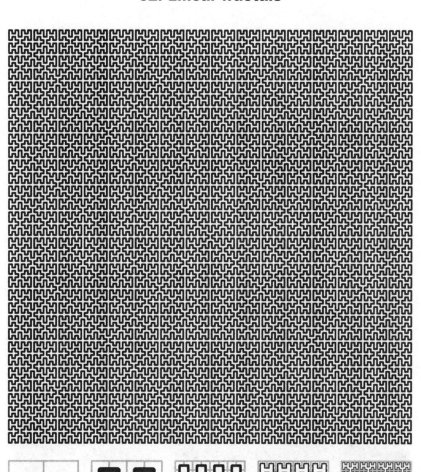

Figure 32-9: "H-curve" method of populating a square grid

These methods can be extended into three or more dimensions to produce curves that fill volumes or higher-dimensionality spaces. The "Newton frame" at the left of Figure 32-8 passes through the centre of every block in a 2×2×2 grid, the shape alongside it uses eight cut-and-crosslinked copies to pass through the centres of a 4×4×4 cube-array. Different versions of Peano-style curves can have different properties, for instance, in how evenly the shape's curvature is distributed, or in how many cuts it generally takes to surgically remove a randomly-selected compact section from the main body of the shape.

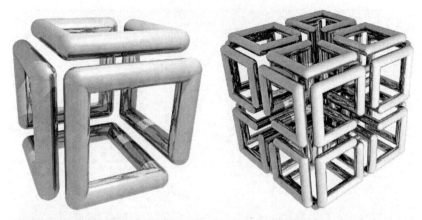

Figure 32-10: Simple and complex edge-following curves

Figure 32-11: Branching 45-degree rotated crosses

33. "Planar" or "Foliar" fractals

"Planar" fractals consist of two-dimensional sections of plane arranged in a higher-dimensional space. The challenge with planar fractals is finding pleasing constructions that don't already correspond to the faces of "solid" fractals.

Figure 33-1: Foliated Squares

Figure 33-2: "Hexflower" fractal

Alt.Fractals

Figure 33-3: Hexflower, with two faces removed

Planar fractals can also suggest new members of other fractal types.

In Figure 33-4 we replace a single equilateral triangle with a star-array of three smaller triangles that are perpendicular to the original shape, but preserve its corners. After about ten iterations, the process has created the substantial planar skeleton of a very definite "solid" fractal, which is then reconstructed from scratch in Figure 33-5. The faces of the new solid show the same basic pattern as Figure 3-19.

33. "Planar" or "Foliar" fractals

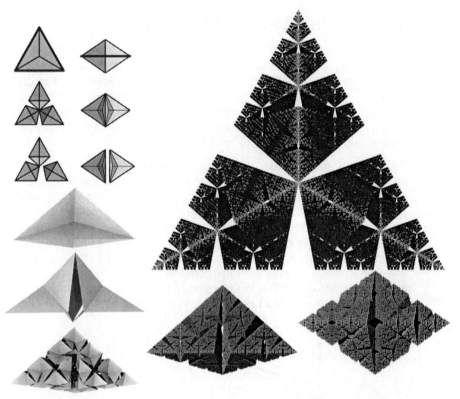

Figure 33-4: A "planar" fractal built from equilateral triangles ...

Figure 33-5: ... and its solid counterpart, the "Delta"

34. "Plotter" Fractals

"Constructional" fractals, made by repeatedly replacing a shape with rescaled, offset, flipped and/or tilted copies of itself, are sometimes referred to as "**IFS**" fractals. This stands for "**Iterated Function System**", where "function" refers to the change between a shape and its offspring. Back when computer hardware was more primitive, the simplest IFS shapes to create were the ones that could be drawn with a computer-controlled pen, or "plotter".

Pen-plotters were often driven by **turtle commands** – sequential instructions that moved a pen or cursor across a page to draw a shape, like a turtle leaving a trail as it crawls across a sandy beach. Turtle commands were crude – the cursor could rotate though a specified angle, advance a specified distance, and maybe lower or raise a pen-nib. Starting with a computer script of "pen" commands, a program could seek out every command that would draw a straight line and replace it with a more complex set of "rotate" and "draw" commands copied from a template, with the resulting script becoming more complex each time the template is was applied.

Templates can be simple or complicated. If a template repeatedly replaces the command, "**draw a line of distance d**" with "**divide d by three, draw a line of distance d, rotate left 60 degrees, draw a line of distance d, rotate right 120 degrees, draw a line of distance d, and rotate left 60 degrees**", then once the template has been applied recursively a few times, the resulting script will make the computer-controlled plotter waltz the outline of the Koch Curve:

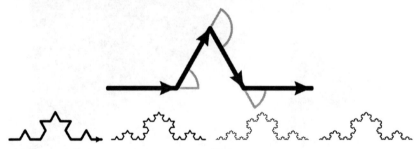

Figure 34-1: "Koch Curve", and IFS Template

Templates can be arbitrary, even based on letters of the alphabet.

34. "Plotter" Fractals

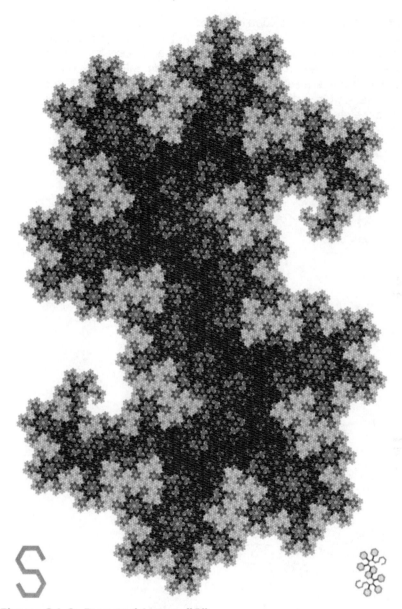

Figure 34-2: Iterated Letter "S"

This image is the result of starting with a single straight line and repeatedly replacing every line-segment with a nine-segment "S" pattern. The darker segments in the resulting hexagonal mesh show where lines have been overdrawn multiple times.

Alt.Fractals

Plotter fractals were ideally suited to computer systems that had limited graphics capabilities and restricted memory. A program could read a target script from one text file, analyse one line at a time, and use its template to write new commands into a second text file, without the entire script having to be in memory at once. The file could then be sent to a plotter that could execute the instructions, like a set of dance steps, to draw the fractal.

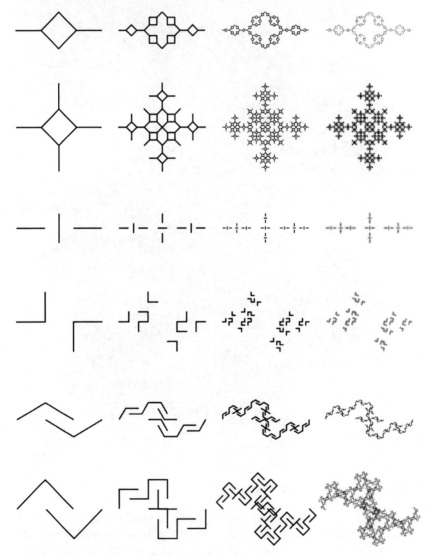

Figure 34-3: A Few Simple IFS Templates

34. "Plotter" Fractals

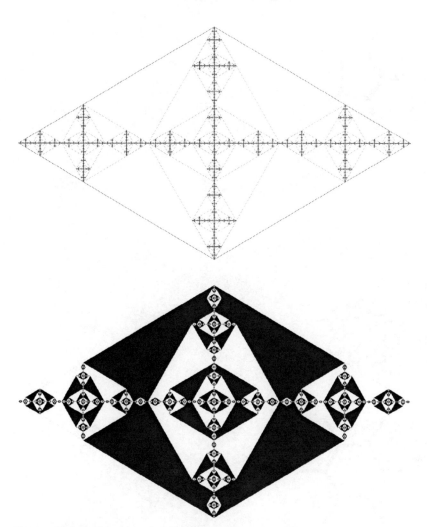

Figure 34-4: Diamond Subdivided

Figure 34-5: Fractal crosshatching

Alt.Fractals

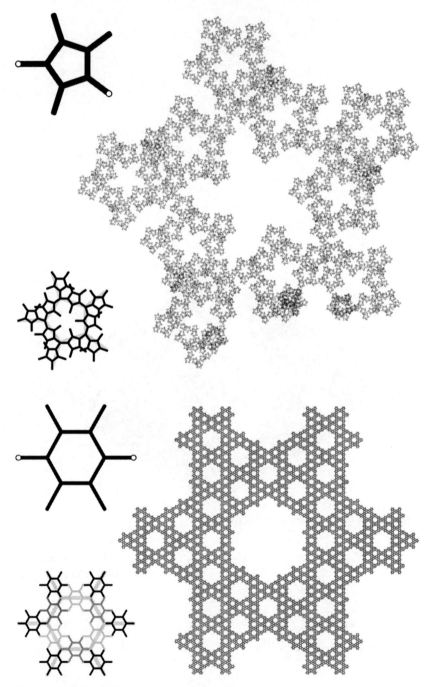

Figure 34-6: IFS Spiked Pentagon and hexagon

34. "Plotter" Fractals

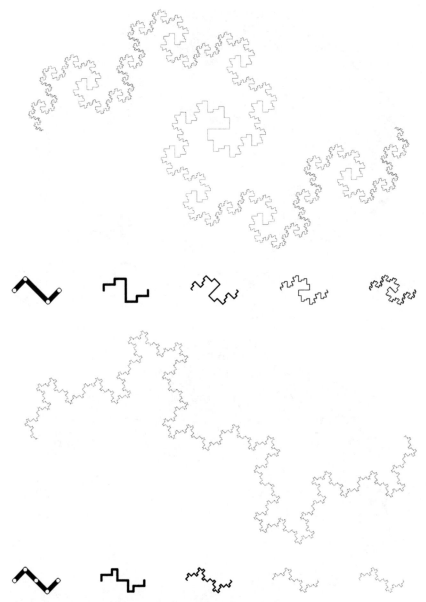

Figure 34-7: Subdividing an IFS template

Although these two templates look similar, the lower template uses four equal-length sections, while the upper uses three sections of two different sizes. The mismatch between the characteristics of its left-turns and right-turns gives the first shape a definite spiral twist.

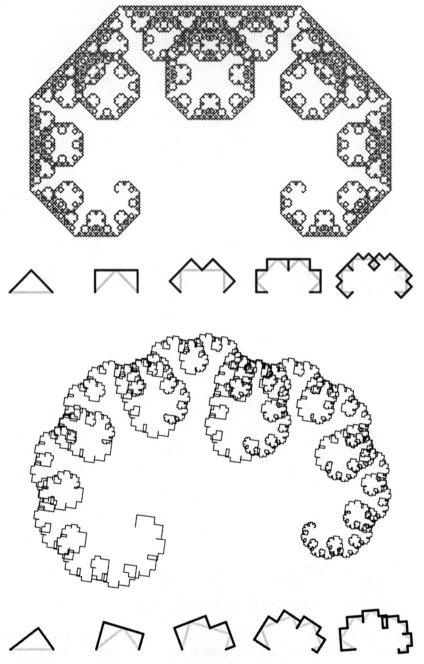

Figure 34-8: IFS Levy Curve and "Brain"

34. "Plotter" Fractals

Figure 34-9: Effect of template alignment

All four fractals in Figure 34-9 appear to use the same template – a simple hexagon with two sides removed – but the alignment of the replacement line-segments can have a dramatic effect on the final shape. While the result of the first template isn't totally unexpected, the "leaf-like" second result is more of a surprise (and is shown in more detail as Figure 35-10). These radical changes in character that can be caused by small changes in approach are one of the factors that have made fractals such a satisfying subject for exploration.

35. Dragons

"Pen-plotter" methods lend themselves to drawing a special class of fractal called a *Dragon*.

Dragons usually consist of a single, continuous, kinked, self-similar line that never crosses itself, and they nest and tile a grid in ways that can seem impossible until you actually try it.

It's traditional to do this using a square grid, but since we're trying to be different, we'll construct one using equilateral triangles instead.

Figure 35-1: "Zagger" Fractal (Iterations 0 to 2)

Dragon fractals often don't look too impressive for the first few iterations, and in this case the shape's fractal character doesn't start to become apparent until iteration five.

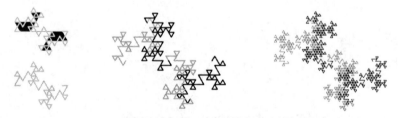

Figure 35-2: "Zagger" Fractal (Iterations 3 to 5)

This particular dragon builds each generation from four angled copies of the previous one. If we're using "turtle" commands, the "zagger" can be drawn by repeatedly replacing every command "**draw a line of length D**" with, "**Divide D by two, turn left 60 degrees, draw a line of length D, turn right 120 degrees, draw a line of length D, draw a line of length D, turn left 120 degrees, , draw a line of length D, turn right 60 degrees**".

35. Dragons

Figure 35-3: "Zagger" Dragons (Iterations 6 & 7)

Figure 35-4: "Zagger Dragon" Star

Figure 35-5: "Zagger" Dragon (Ninth Iteration)

Figure 35-4 (previous page) has six Zagger Dragons meeting at a common point.

Since every "void" in a dragon can be reached from one side, dragons can intersect and interlock in surprising ways without their lines actually crossing.

35. Dragons

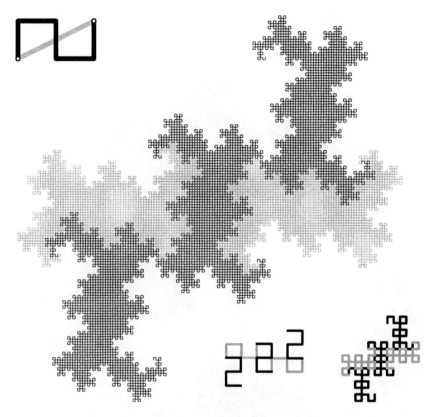

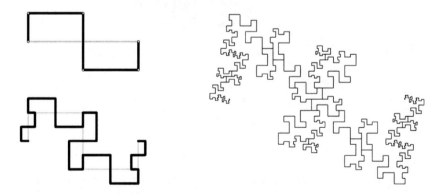

Figure 35-6: A more traditional "Square-Grid" Dragon

Figure 35-7: A more complex dragonlike shape with two scales

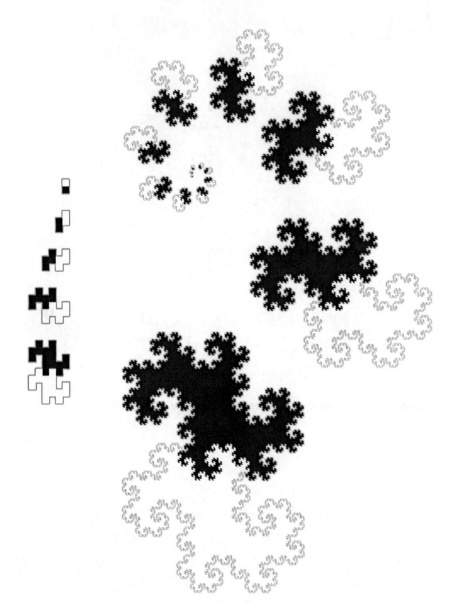

Figure 35-8: Standard Block "Dual Dragon" Construction

35. Dragons

Figure 35-9: "Dual Dragon" Spiral

The ability of dragons to tile different generations of shape together can be used to build more complex shapes than just the simple "islands" of section 31. Figure 35-9 is a single spiral tiling of different generations of the "Dual Dragon". More complicated multi-armed spiral tilings are also possible.

Figure 35-10: A "Broken" Dragon

The slightly oak-leafy fractal in Figure 35-10 fulfils most of the criteria for being a dragon … it tiles with itself and with its subcomponents, and it doesn't self-cross … but since it consists of a cascading series of distinct islands, it fails the last condition, of being able to be drawn with one single continuous line.

This shape's template is given in Figure 34-9.

36. Fractal Plantlife

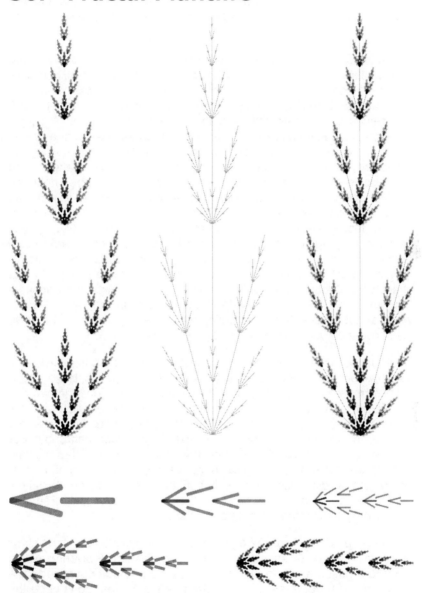

Figure 36-1: A Simple IFS Grass

The "grass" diagram in Figure 36-1 is made by repeatedly cutting a line into two equal parts, and tilting the base section and a copy by +/- fifteen degrees. It looks like a flowering grass, because the binary sequence produces identically-sized clusters regardless of where they appear on the network. A grass plant in flower needs to produce a large number of seeds of a fixed size, irrespective of how many times the parent stalk branches.

The examples in this section have been chosen for simplicity rather than accuracy – "proper" computer-generated fractal plantlife usually involves more sophisticated methods than the simple 45-degree branching and fixed geometrical ratios shown here. In the real world, proportions and angles shift with the number of iterations. A supple tree may start with narrower joint angles that spread as the tree grows, giving a variation from wide to narrow angle as we follow the branching. A head of flowers might start with narrower angles and progressively "splay". Many plants have alternating staggered right and left branchings to avoid too many intersections at a single region, and we also have to take into account drooping due to gravity, and different scaling factors applying for different parts of a plant. Our "grass" example may apply a factor-of-two scaling to the seed regions but a different scaling for the main stems. Hormone variations across a growing plant produce cut-offs and inhibition factors, as do the shadows that leaves cast on other leaves. Root growth often appears random, but shows proximity inhibition – root growth often "stunts" when a root senses that a competing root is nearby. And of course, we also have a certain amount of accidental random variation.

Real leaves have branching patterns of ribs that don't intersect, with potential conflicts resolved locally. New branch growth tends to be inhibited near existing branches. The distinctive variations that we see in different species are partly a function of inbuilt sets of rules, abut are also a product of how the physical results of those rules interact in the real world. Humans normally have five fingers per hand, but our templates don't explicitly contain the number five. They contain instructions for growing fingers in sequence, in a way that *usually* peters out at digit number five, but if these systems are disrupted, people can be born with different numbers of fingers or toes. Features can be epigenetic rather than hard-coded.

"Fern" shapes are a special case, because simpler plants don't show the same degree of specialised scale-optimisation. Flowering plants need to produce and bear seeds and flowers and fruit optimised for a particular size, but ferns, which reproduce via spores, don't need the same scale-specific features. They can produce simpler general-purpose rescalable structures that can be convincingly replicated using fractal techniques, to the extent that the outlines can sometimes be indistinguishable from those of real plants. Reproducing the

36. Fractal Plantlife

characteristics of the higher plants (such as a leaf's veining patterns) usually involves more artistic "tweeking".

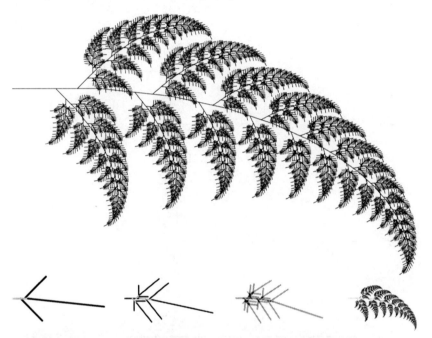

Figure 36-2: A "fractal fern" image, and template

Alt.Fractals

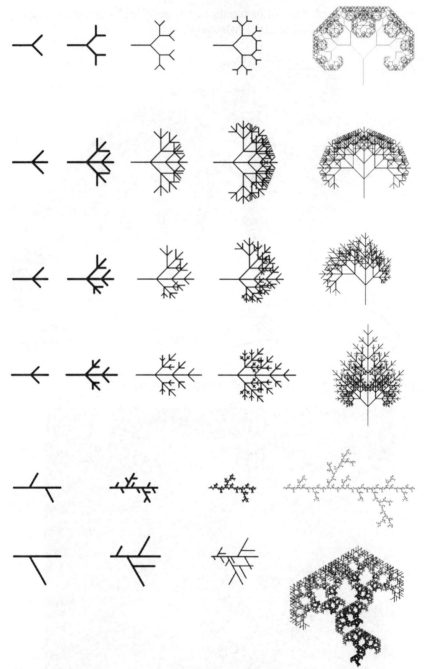

Figure 36-3: Some basic "tree" templates

36. Fractal Plantlife

Figure 36-4: A simple three-branch tree

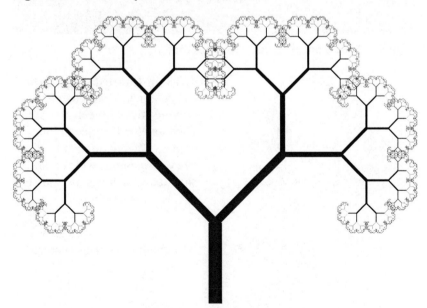

Figure 36-5: A simple two-branch tree

Figure 36-6: Romanesco "cauliflower"

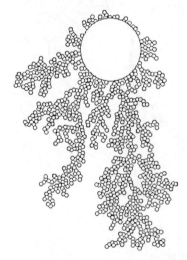

Diffusion-Limited Aggregation (DLA) fractals can be made by allowing sticky objects to collide with a "seed" object, along random zig-zagging paths. The resulting outgrowths create shielded channels and voids in the structure.

DLA can be used to grow branching root-like structures.

Figure 36-7: Diffusion-Limited Aggregation (DLA)

37. Advanced IFS

The "fern" exercise in the previous section shows that the simplest IFS techniques often aren't able to draw the shapes that we want. Trying to plot the fern threw up three practical problems:

1. There was no simple relationship between a feature's iteration stage and its size

2. Cumulative errors in the tiniest parts of the shape could cause subsequent misalignments of large branches

3. The "stem" regions of the fern didn't branch.

We can solve the first problem by using more iterations than we need, but adding a "cutoff" condition that prevents further division below a particular size threshold. This avoids wasting time plotting detail that's too small to see, while larger regions remain sparsely populated.

The second problem is solved by saving the cursor's current values whenever a template is launched, and restoring them after the template (and all its subtemplate iterations) have finished drawing. This lets information propagate "downwards" through the branched fractal, without any errors travelling back up through the network. IFS languages typically have special "register" commands for this job, that can store and restore the cursor's state.

Tackling the third problem requires more sophisticated templates and/or multiple templates. The "fern" template has two types of line segment: one that branches normally, and another (for the stem links) that draws a line once, and then either stops, or overdraws the same line repeatedly. "Template A" calls both templates, A and B, but "Template B" (for the stems) only calls itself.

Figure 37-1: Modified IFS patterns

Alt.Fractals

There are some legitimate cases where we might need multiple templates: if we want to solidly tile space using Sierpinski pyramids and octahedra (Figure 10-3), we need the void within each pyramid to trigger the octahedral template, and the cutouts in each octahedral template in turn to call the pyramid routine. In the triangular shape in Figure 4-8, the triangle template calls three smaller triangles and three wedges. The wedge template in turn calls further wedges and two triangles.

For drawing realistic-looking fractal plants, two templates often aren't enough. Figure 36-2 draws a decent single fern frond, but real ferns usually have a number of distinct compound leaves arranged in a spiral around a central body: a helical template calls the leaf template, which in turn calls the branching and non-branching templates. With more complex plants we might have separate patterns for twigs and leaves and branches and flowers and roots, and at least some of our templates will have to be three-dimensional to allow for helical branching patterns that don't exist in a flat plane.

Figure 37-2: A "Crop Circle" fractal

Complex templates also allow special effects and gimmicks: to make the "crop circle" fractal in Figure 37-2 we start with a "fern" program, but modify it so that the coordinates of each final non-branching segment are used to draw a circle instead of a line. To hide its "ferny" origins, the template's also been given rotational symmetry – the result is something that still looks somehow organic, but alien.

38. Number

Integer-based number systems are designed to be fractal: the mechanics of the system work exactly the same at the tenth decimal place as at the millionth. They are iterative systems that try to get as close as possible to any real number by progressively iterating more and more digits. If we try to write "one third" in Base Ten, the iteration (0.3333333333333'...) goes on forever without ever reaching its target. For irrational numbers (like Pi and the square root of two), the target is unreachable using conventional integer iteration, regardless of which base we use.

Binary	Base 3	Base 4	Decimal
0000	0000	0000	0000
0001	0001	0001	0001
0010	0002	0002	0002
0011	0010	0003	0003
0100	0011	0010	0004
0101	0012	0011	0005
0110	0020	0012	0006
0111	0021	0013	0007
1000	0022	0020	0008

One of the ways that we can express the self-similar properties of a conventional number system is with a "$z=xy$" graph.

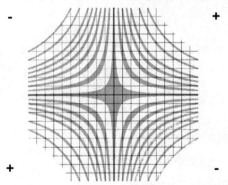

This gives a three-dimensional saddle-shaped surface. No matter how far we zoom in on the centre of the curve, we always get an exact copy of the shape we started with. The surface is scale-neutral: draw a horizontal contour across the surface and it can be used to represent a value of one, or a million, or Pi. To assign integer labels to the number-surface is then a matter of quantisation – but the surface exists independently of any integers we might choose to attach to it.

Alt.Fractals

Figure 38-1: Base Two branching (strip diagram)

Figure 38-2: Binary Squares

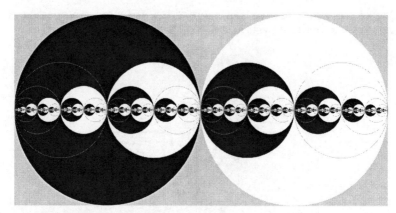

Figure 38-3: Binary Circles

Figure 38-4: Base Three (strip diagram)

38. Number

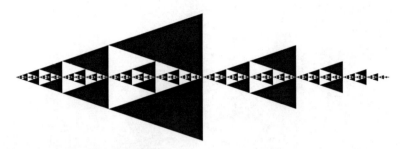

Figure 38-5: Binary Diamonds

Figure 38-6: Binary Chessboards

Figure 38-7: Derived pattern

Figure 38-8: A Base Three Numberspace

The fractal properties of a number system can sometimes be directly exploited to plot shapes.

One of the differences between the Mandelbrot/Julia type fractals and the earlier shapes in this book is that the Mandelbrot Set lets us "colour in" a single point in space, independently, without having to first draw the entire surrounding shape, where the "constructional" fractals like the Sierpinski Triangle are usually considered to be things that have to be drawn or built before we can find what appears at any given point.

But some fractal shapes can be produced with either method. To construct the Menger Sponge, we can divide a cube into a 3×3×3 grid, mark the centre blocks on each axis, and delete any blocks that get counted two or more times. To assign row, column and layer values corresponding to x, y and z coordinates only needs three possible values per axis, per iteration. If we specify the x, y and z coordinates of a point in space within the top-level block using numbers in Base 3, the first digit of each coordinate selects the first iteration block, the second selects the second-iteration sub-blocks, the third digit refers to the third-iteration divisions, and so on.

38. Number

Figure 38-9: A "point-tested" slice through Figure 38-11

All we then have to do to check whether a specified point refers to "solid" or "space" is to read off the digits and compare. Choose a desired number of iterations, "**n**", crop the "Base 3" *x*, *y* and *z* coordinates to **n** places, then compare the three sets of results against each other, digit by digit. The "middle" value of "1" refers to a central column or row: if two or more coordinates have a "1" in the same digit- position, it means that the point must lie within a void, otherwise the point is "solid" for the first "**d**" iterations.

This method lets us plot images of Menger-like solids (complete or partial) in three or more dimensions, and was used to plot the Menger-related images and cross-sections ("flat" and distorted) shown in this section and in section 39.

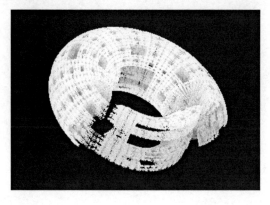

189

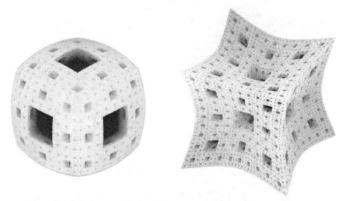

Figure 38-10: Distorted Sponges, using a coordinate transform and point-testing method

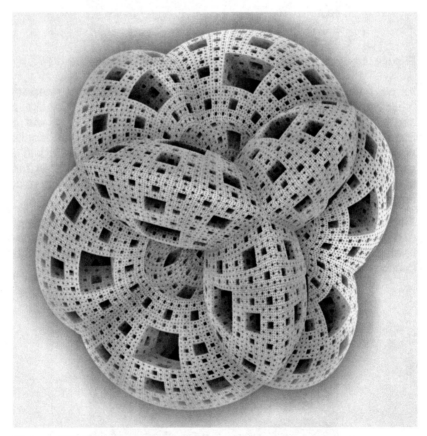

Figure 38-11: The Menger Sponge twisted inside-out

38. Number

Figure 38-12: 3D Cross-Section through a 4D Hypersponge

Point-testing methods can map solids in more than three dimensions. Figure 38-12 plots a solid cross-section cut at an angle through the four-dimensional counterpart of the Menger Sponge. In the 4D sponge, block-shaped struts connect the eight corners of "start" and "end" Menger Sponges, so when we cut the 4D sponge across its middle, we expect to see eight connections. Cutting it across the "corner-on" view gives 2D and 3D shadows that are more difficult to visualise. With point-testing we can simply feed in the coordinates and plot the result, which in this case turns out to be a fractal based on eight corner-cubes that form a larger cube (we can only see seven in the plot because cubes #7 and #8 both lines up with the centre of the view). This is simple unconnected fractal " cube dust", but it would be difficult to prove this with conventional modelling software that only worked in three dimensions.

39. Inside the Menger Sponge

Slicing the Menger Sponge at right angles to its main axes gives a fairly obvious sequence of square islands and Sierpinski carpet derivatives.

Cutting it at right angles to a line running *between two opposing corners* gives a rather different set of fractal shapes based on hexagons, diamonds, stars and triangles.

Figure 39-1: Nine Slices through the Menger Sponge

39. Inside the Menger Sponge

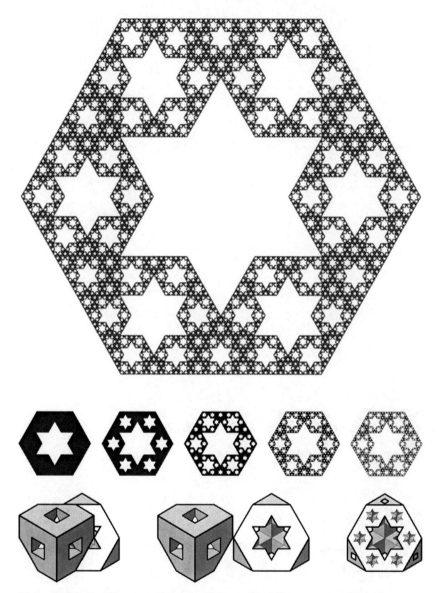

Figure 39-2: Menger Sponge "Zero Star" cross-section

The outline of a cube, viewed corner-on, is hexagonal. Cutting the cube through its centre, perpendicular to this view, gives a smaller hexagon as the "cut" face. The cube-shaped void at the centre of the Menger Sponge makes a hexagonal cutout, and parts of its three intersecting shafts then turn the centre hole into a star-shape. Symmetry then forces all the other holes to be star-shaped, too.

Figure 39-3: Menger Sponge cross-section

39. Inside the Menger Sponge

Figure 39-4: "Mitsubishi" Menger Sponge cross-section

Sometimes the construction of a fractal that seems purely decorative turns out to have a deeper geometrical significance.

Here, a "Mitsubishi" ("three diamonds") fractal from Figure 4-7 turns out to be another angled cross-section taken from the Menger Sponge.

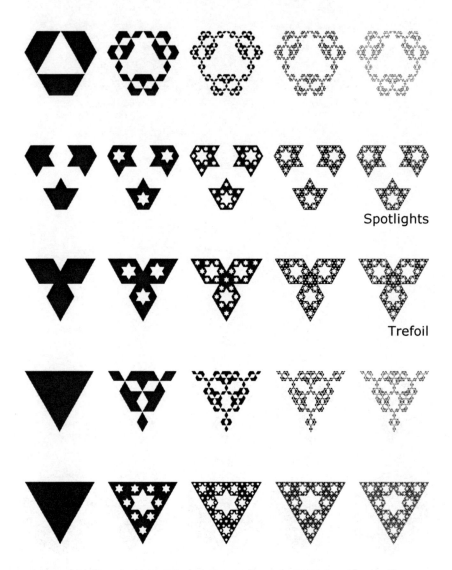

Spotlights

Trefoil

Figure 39-5: A few more Menger Sponge slices (resized)

40. The L-block

The starting-shape in Figure 40-1 is made from four cubes connected together to give a "triple-L".

Since the solid has a volume of "four", and its enclosing cube has a volume of 2×2×2 = "eight", it fills exactly 50% of a cube-shaped container. It will interlock with a second rotated copy it itself, like a three-dimensional Yin-Yang symbol, and completely fill the bounding cube. The **L-block** is a space-filling solid.

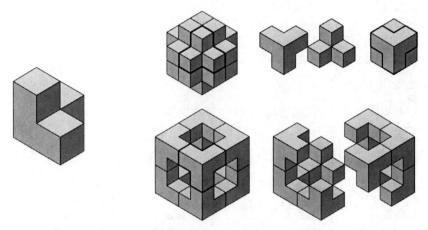

Figure 40-1: Some L-block tiling modes

The solid can be stacked or subdivided fractally to produce further space-filling solids and arrays.

The simplest method is to replace each of the four cube-shaped units with a smaller copy, with every duplicate having same orientation.

In two dimensions, this gives a nested series of "L" shapes, and the repeated subdivision eventually shears away everything in the diagonal top-right half of the figure, leaving the set of stacked right-angled triangles in Figure 40-2.

Figure 40-2: "Sierpinski-like" right-angled triangle

Since we want to do this in three dimensions rather than two, we get a "cubist" version of the Sierpinski Pyramid, with Figure 40-2 on three of its four sides (Figure 40-3). The fourth side is "stepped" and has threefold rotational symmetry. As the number of iterations increases and the step-size gets smaller, it gets ever-closer to being a "proper" Sierpinski triangle.

40. The L-block

Figure 40-3: "Sierpinski-like" right-angled pyramid

There are other ways to stack the block. Adding a reflected copy to each prong gives us the 3D "library fractal" of Figure 11-5, and completing the sequence to produce a fully closed cube gives the "1, 0, 0, 1" variation on the Menger Sponge in Figure 9-1. The complementary t shapes in Figure 40-1 both have 32 filled blocks within a 4×4×4 = 64-block grid, and their 50% fill ratio is still valid when we stack more of our blocks together without overlaps.

Figure 40-4: Dual-Interleaved Space-filling Array

Alt.Fractals

The "stretched Menger sponge" in Figure 9-1 fills space. Imagine stacking a three-dimensional array of the basic first-generation building block to produce an infinite grid. The edge-struts of four adjacent blocks will fuse together to give a set of thicker struts that are two units wide, surrounding a similar network of holes that are also two units wide. The first-iteration version of the sponge has a 50% density, and two identical infinite arrays of the sponge will interleave to completely fill space (Figure 40-4).

If we now punch the next iteration of holes into the solid array, we find that this again halves the volume of each array, and of each individual block, and again produces networks of holes and remaining material that are again inverse copies of each other. We can keep going like this indefinitely, after a thousand iterations, we can have 2^1000 interleaved copies of the shape occupying the same space.

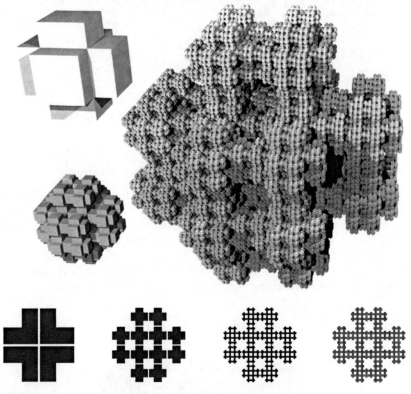

Figure 40-5: "Triple-Cross" fractal, and shadows

The inverse of the sponge building-block can also be built from eight copies of the L-block, giving the "triple-cross" solid in Figure 40-5.

41. Fractal Circles

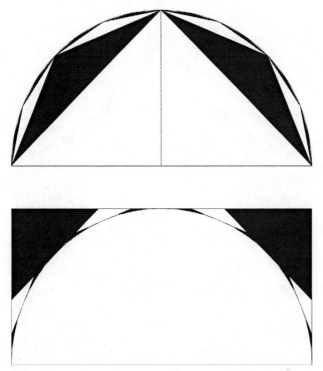

Figure 41-1: Successive Approximation of a Circle (working outwards and inwards)

One of the classic methods of trying to calculate the area of a circle is by considering it as an infinite-sided regular polygon. Starting with a square, we successively add or subtract triangles to double the resulting shape's number of faces at each generation.

We can choose to work outwards (continually replacing each face with a pair of smaller faces), or inwards (progressively lopping off the shape's corners). Using both methods together, we can obtain a pair of "high" and "low" estimates that converge on the true, intermediate value.

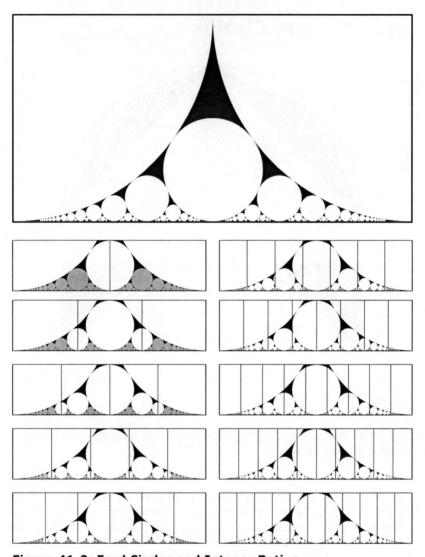

Figure 41-2: Ford Circles and Integer Ratios

Ford Circles appear as part of a packing pattern when we try to tile the space between a pair of circles and a straight line. They have two surprising features: (1) the points where they touch the line form a regular integer-division grid, and (2) if we iterate the grid to find the position of each new generation of circles, the new circles added at each stage somehow seem to manage to be the same size as each other, even though their respective parents' sizes are often different.

41. Fractal Circles

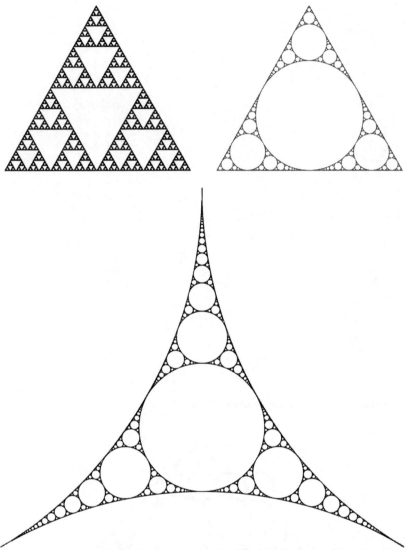

Figure 41-3: Circled Counterparts of the Sierpinski Triangle

Ford Circles can be thought of as a special case of packing a region bounded by three circles, where one of the circles is infinitely large (giving an effectively straight line).

Circle-within-circle packing is an old problem that was worked on by **Gottfried Leibniz**, **René Descartes**, **Lester Ford** and **Frederick Soddy**, amongst others, so topologically, the Sierpinski Triangle is a simplified, stylised map of a much older problem.

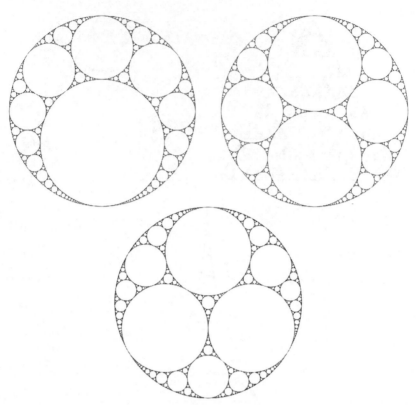

Figure 41-4: Circular "Gasket" fractals

42. Circular-faced fractal solids

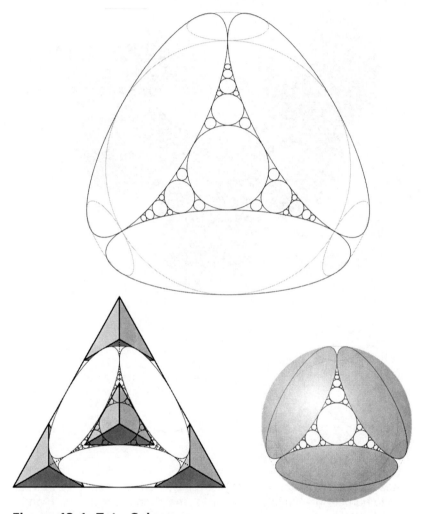

Figure 42-1: TetraSphere

The "net" examples in Figure 41-4 can be stretched over the surface of a sphere to describe a family of fractal solids. We've already met a few shapes that were intermediate between members of the family of five regular solids (such as the truncated octahedron in Figure 8-3, which was the offspring of an octahedron and a cube). For these "new" fractally-faceted shapes, we treat the sphere as if it is a sixth regular solid that can be combined with any of the other five.

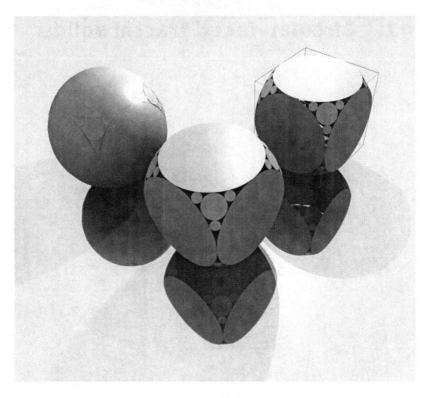

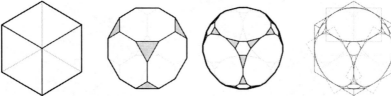

Figure 42-2: CubeSphere

Since a sphere can (in a sense) be considered as being built from an infinite number of infinitely small faces, the offspring of a sphere and a regular solid has to have an infinite number of faces, too.

The resulting shapes are fractally-faceted, with each face becoming a perfect circle after an infinite number of iterations.

We can construct versions of these shapes around any of the five regular polyhedra, and also some of their "semiregular" relatives. The results of trying to fractally-facet an *irregular* solid are less predictable.

42. Circular-faced fractal solids

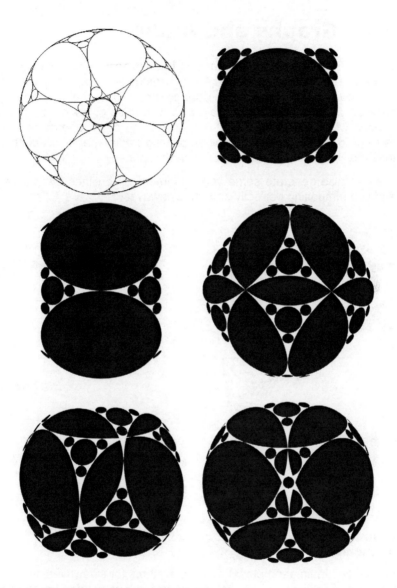

Figure 42-3: CubeSphere Projections

43. Graphs and Audio

Linear IFS fractals can allow parts to be displaced in any direction so that a shape "overhangs" itself. A more restricted form of linear fractal only allows points on a line, surface or solid to be displaced or sheared at right angles to a main axis. We raise or lower sections of surface when modelling fractal mountains and other landscapes, and raise or lower sections of line when plotting fractal graphs of functions that very with time, such as stock market reports.

We can also generate some very "geometrical" outlines by "packing" and superimposing perfectly-smooth sinewaves.

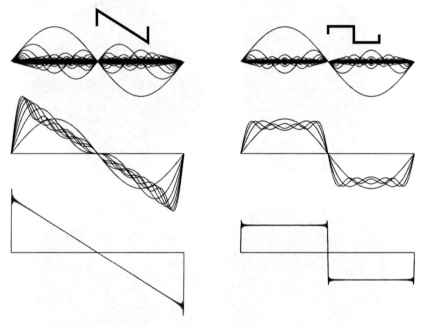

Figure 43-1: "Ramp" and "Square" waves, first 128 harmonics

For the two main geometrical packing systems for sinewaves, we want the offspring copies to have the same proportions as their parents (for audio, we'd say that "amplitude is inversely proportional to frequency", or A="1/f"). With every integer harmonic, the result tends towards an infinitely-close approximation of a ramp wave (left hand figure). If we only use the odd-numbered harmonics (we skip the multiples of two, four, eight, etc.) the summed sinewaves converge on a simple symmetrical rectangle, or square-wave (right hand figure).

44.   Rope and Braiding

Figure 44-1: Rope iterations

Rope and cord have fractal structures. Individual fibres are wound or braided to produce thicker composite structures that are in turn wound or braided together to give progressively larger (and stronger) ropes. For standard wound ropes, the winding direction is alternated at each stage – this guarantees that the fibres can't be made to "splay" by forcibly untwisting a section of the rope. Figure 44-1 is calculated for four generations of windings.

In wound ropes, fibres and clusters orbit common centres. In braids, the stands weave and wind around each other but repeatedly exchange partners, "barn dance"-style.

45. Villarceau coils and Toroidal Packings

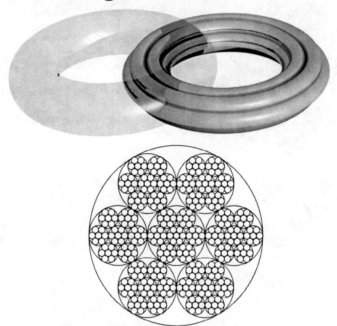

Figure 45-1: Trivial torus-packing

Some methods of packing toroidal solids are more obvious than others. Stacking rings within rings "in parallel" so that they all circle the same parent torus mouth (Figure 45-1) is fairly straightforward. The component rings at any iteration can have the same *minor* radius but rings have multiple major radii and proportions, and the larger pattern is only a spun extrusion of a two-dimensional pattern.

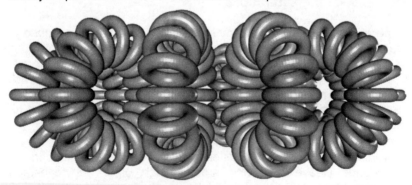

Figure 45-2: Torus iterations

45. Villarceau coils and Toroidal Packings

The two methods can be mixed, but a third, more sophisticated approach is to loosely fill the torus with tilted rings that each circle the torus limb *and* the torus mouth.

The theory of circular toroidal cross-sections was worked out by **Antoine-Joseph Yvon Villarceau**, who pointed out that for any given simple torus, there's a cutting-angle that creates pairs of interlinked circles as a cross-section. If we call the major radius of the torus "**R**" and the minor radius (giving the "fatness" of the torus limb) "**r**" then the full range of possible torus proportions runs from **r=0** (a simple circle) to **r=R** (a torus so fat that its central passage has shrunk to a point). The formula for the "special" angle for tilted circles drawn on a torus is just **Tan A = r/R** .

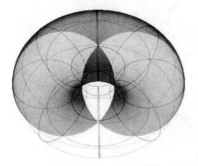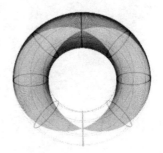

Figure 45-3: Villarceau Circles

Villarceau circles all have the same major radius as their parent torus, so the continuum of possible doughnut surfaces that can be drawn with a given fixed size of Villarceau circle will all "nest" together. Expanding each circle to a finite-volume torus gives a packing method for creating a series of toroidal shells ("**Villarceau coils**") that all fit inside one master torus, and which all have component rings of exactly the same size.

Each ring can be replaced by another Villarceau coil of thinner rings, again with the same major radius, which can in turn be approximated by Villarceau coils of even thinner rings.

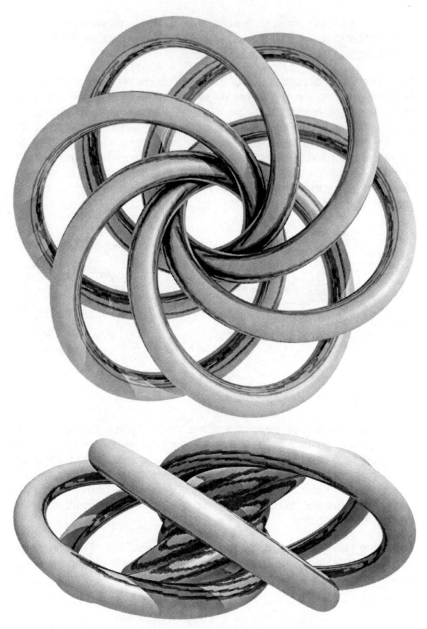

Figure 45-4: A Villarceau Coil of interlinked rings

45. Villarceau coils and Toroidal Packings

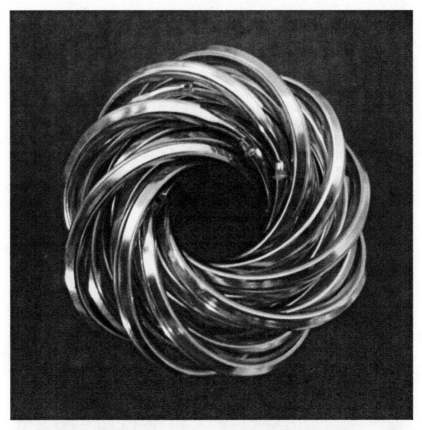

Figure 45-5: Nested Villarceau Coils (using keyrings)

Figure 45-6: Villarceau Coil packing sequence

Each ring is equivalent to a wound coil whose "winding ratio" (number of winding per complete circuit) is exactly "one". Although the rings in a single Villarceau coil (or set of nesting coils) are topologically trapped, any pair of rings is free to migrate through the stack and exchange positions, as long as all the rings share the same tilt polarity.

46. Kaleidoscopes and Mirrorspace

When two mirrors directly face each other, the view for someone standing between them shows an apparent infinite tunnel of parallel-reflected mirrors within mirrors extending in both directions (this configuration is known as an **infinity mirror**). With three or more mirrors we can surround a section of plane on all sides, giving a **kaleidoscope** effect. Three mirrors produces the appearance of an infinite plane tiled by identical triangular sections, and with four mirrors, the infinite grid seems to be made of squares. Since each reflection includes any branchings of the previous one, we can quickly get an extremely complex image.

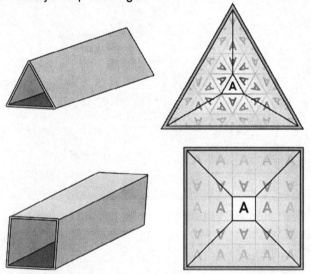

Figure 46-1: Traditional three-mirror and four-mirror Kaleidoscopes, and internal views

We can also use mirror-arrays to create recursive grids in **mirrorspace** that would be impossible in normal space.

Three *curved* mirrors can create a virtual space in which the number of triangles meeting at each point is more than or less than six. Figure 46-2 (facing) shows the view inside a curved three-mirror kaleidoscope with corner angles of 45 degrees, with eight of the triangular units touching at each corner-point in mirrorspace. When an object is put between the mirrors, the reflections show an infinite array of copies of the object, clustered in rings of eight in a way that can't be achieved in a standard 3D volume.

46. Kaleidoscopes and Mirrorspace

Figure 46-2: "Eight by 45" Triangle-space: empty and populated

The new plane fits more space into itself that would normally be possible in two dimensions, and the resulting virtual space arguably has some fractal properties.

We can also extend the principle into three dimensions by taking a regular twenty-sided dodecahedron and teasing out its corners, stretching them from the centre until each (curved) pentagonal face has five corners of 90 degrees rather than 108. The "spiky" dodecahedra can "tile" mirrorspace, with eight cells meeting at each shared corner.

Alt.Fractals

In Figure 46-2, the view in the thee convex mirrors is of something that mathematicians call a **negatively-curved space**. Figure 46-4 shows the opposite result from using three concave mirrors. This gives a view of a space with **positive curvature** that can be tiled using a finite number of cells.

Figure 46-3: Curved Kaleidoscopes with 45°, 120° and 90° angles

In this case, the mirrors are curved to intersect at 90 degrees, four cells appear to meet at each corner, and the entire plane appears to be tiled with just eight cells.

Figure 46-4: Internal reflections from Figure 46-3

Placing a ball in the space between the mirrors gives a reflected universe containing eight balls, the most distant eighth ball seeming to completely surround the observer.

46. Kaleidoscopes and Mirrorspace

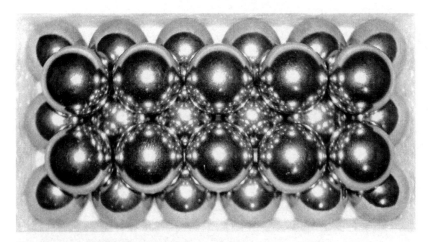

Figure 46-5: Reflective ball-bearings

An easier way to see the result of infinite recursive grids is to look at a pile of polished ball-bearings.

Each mirrored sphere shows a complete 360-degree reflection of its exterior – it appears to contain the entire outside universe reflected and rescaled inside its volume. That distorted mirror-universe contains images of the similar views of the surrounding mirror-balls, and the recursive effect is infinite.

Figure 46-6: Mirrored spheres stacked in two dimensions

Figure 46-7: Mirrored spheres stacked in three dimensions

The geometry of negative-curvature spaces was worked on by **Eugenio Beltrami** in the 1860's. Nearly half a century later, **H.S.M. ("Donald") Coxeter** published a disk diagram of negative-curvature space, and helped to inspire **M.C. Esher**'s famous "Circle Limit" series of woodcuts.

47.   Life is Fractal

Nature is fractal. It can be difficult to distinguish between pictures of writhing shapes carved into beach sand by rivulets of water over a region of a few metres, and the larger-scale tracks of rivers covering tens or hundreds of kilometres. Fluffy clouds have self-similarity at different scales, lightning produces critical branching patterns, and it's easy to produce convincingly recognisable simulations of jagged mountain profiles using fractal software even though our usual mathematical vocabulary doesn't have words that adequately describe what those shapes are.

Life is fractal. Coral growths branch fractally to give the biggest catchment area for a given mass, geckos' toes have fractal hairs that let them cling to surfaces by atomic forces, and plants have developed branching structures to maximise the surface area for photosynthesis at minimum structural cost. Inside our bodies we have branching circulatory and nervous systems, and the pattern of airspaces in our lungs is fractal, too. This doesn't just maximise surface area for gas exchange, it allows complex systems to be grown from simple recursive rules, without any sort of externally-imposed design.

Populations spread fractally in space and time. Our lives form lines in spacetime that kink and twist with our movements and touch against those of others. In four dimensions, a family tree represents a continuous repeatedly-branching chain that at larger scales becomes the branching of populations, and then of entire species and phyla in the evolutionary tree.

History is fractal. History repeats itself ... but not exactly. Themes and patterns emerge. Those who don't learn the lessons of history may be doomed to repeat it, but with variations. In **Jorge Luis Borges**' "*The Garden of Forking Paths*", a character's will leaves a labyrinth and a book, the book apparently existing as a set of multiple drafts. The book is the labyrinth: the choices we make represent a series of decision-forks in a wider dynamic fractal web of possibilities.

Society is Fractal. Twitter is fractal. Information fed into Twitter propagates and spreads and mutates, or dies out, leaving a trace of forking paths. Social media networks are fractal, as are telecommunication networks, with their trunkline and local branches. Our road network systems are fractal with A-roads, B-roads, and a set of low-level tracks and alleyways functioning like a body's arteries and capillaries.

219

Alt.Fractals

Physics also appears to be fractal, although it would seem to take the form of an "atomistic" fractal, with a Hausdorff dimension that varies with scale.

Even cosmology might be fractal. If the physics of our universe is defined by a handful of critical "seed" parameters that generate the entire interior structure, then our universe might be considered as being one point in a multidimensional parameter-space. Our observable universe might exist as a horizon-bounded lobe on a larger surface that in turn exists as a lobe on something even bigger, with each lobe inheriting some parameter ratios from its parent, but having its own unique local physics and local fundamental constants.

We used to view the universe as if it was a completed piece of clockwork, with fixed rules dictated by polite mathematics and classical geometry, and any anomalous "pathological" equations being the unwanted exception. Nowadays the view has shifted: instability and chaos appear to be the norm, with chaotic orbits being the universe's way of reaching out to explore and embrace a wider parameter-space. Things that we think of as fixed properties may be emergent characteristics of an underlying system that's inherently chaotic. The universe isn't stable, its size isn't fixed, and it's that evolution of physical states that has allowed it to produce life-sustaining regions rather being a simple sterile void. Fundamental physics theory isn't totally fractal *yet*, but quantum mechanics coupled with the concept of atomistic fractals may eventually get us there.

Meanwhile, we get to look at the pretty pictures.

48.   Table of Figures

Alt.Fractals

48. Table of Figures

48. Table of Figures

Lightning Source UK Ltd.
Milton Keynes UK
26 November 2010

163439UK00001B/6/P